WHAT PEOPLE ARE

Primal A\

Rob Wildwood's fascinating spiritual odyssey around the world has finally manifested for us all to read. His special gift in communicating with the 'spirit of place' has given us a vital message in how we can re-connect with the ancient wisdom and more importantly, integrate this knowledge, so that we can create a new paradigm for the future.
Gary Bitcliffe, dowser and author of *The Spine of Albion*

This beautifully written book explains how we as a species have progressively forgotten the importance of our relationship with Mother Earth. It will reawaken within you dormant soul memories and re-establish a fervent desire to be at one with the trees, the rocks, the water and their spirits once again. It's a book that calls you home.
Alphedia Arara (formerly **Fiona Murray**), channel, spiritual workshop facilitator, environmental scientist and author of *Messages from Natures Guardians*

At last! A truthful, frank and well researched look into how we, as a human race, were created from a dream of love and respect, to how we have become shackled by fear and corruption. It then offers us a deep understanding that we have the power to wake up from the manipulative nightmare and dream a new dream, one which connects us into our healing and brings us back to who we really are and what we as a species have come here to do!
Jay Oakwood, director of The Bridget Healing Centre, The Call of the Shaman Training and The Healing Drum, Glastonbury UK

Primal Awareness is a beautifully written treasury of insights that

uncovers history's hidden truths! A must read for anyone who wishes to reclaim their spirit and raise their consciousness to bring themselves, and this world, back into balance.
Flavia Kate Peters, author of *Shaman Pathways – Way of the Faery Shaman*

Primal Awareness
Reconnecting with the Spirits of Nature

A short history of mankind's separation
from nature and the world of spirit
...and what we can do to reconnect

Primal Awareness
Reconnecting with the Spirits of Nature

A short history of mankind's separation
from nature and the world of spirit
...and what we can do to reconnect

Rob Wildwood

Winchester, UK
Washington, USA

First published by Moon Books, 2018
Moon Books is an imprint of John Hunt Publishing Ltd., Laurel House, Station Approach,
Alresford, Hants, SO24 9JH, UK
office1@jhpbooks.net
www.johnhuntpublishing.com
www.moon-books.net

For distributor details and how to order please visit the 'Ordering' section on our website.

Text copyright: Rob Wildwood 2016

ISBN: 978 1 78535 656 8
978 1 78535 657 5 (ebook)
Library of Congress Control Number: 2017931222

All rights reserved. Except for brief quotations in critical articles or reviews, no part of this book may be reproduced in any manner without prior written permission from the publishers.

The rights of Rob Wildwood as author have been asserted in accordance with the Copyright, Designs and Patents Act 1988.

A CIP catalogue record for this book is available from the British Library.

Design: Stuart Davies

Printed and bound by CPI Group (UK) Ltd, Croydon, CR0 4YY, UK

We operate a distinctive and ethical publishing philosophy in all areas of our business, from our global network of authors to production and worldwide distribution.

CONTENTS

Part I – The World of Spirits   3
The Shaman's Voice

Part II – 19 Levels of Separation   11
The Naming of the Spirits   13
Stories & Myths   19
Shamanism   22
Altering our Environment   26
Settlement and Farming   32
Tribal Culture and Division of Labour   38
Possessions, Wealth and Status   41
Organised Religion   44
Warfare, Dominance and Patriarchy   48
Cities, States and Bureaucracy   51
Writing   54
Monotheism   58
Reformation   67
Natural Philosophy and Science   70
Industrialisation   77
Colonialism   82
Mechanistic Thinking   85
Economics and Capitalism   93
Modern Technology   98
Summary – The World Today   101

Part III – The Future   107
Reconnecting with our Indigenous Ancestors   109
The Return to a New Age   112
Unconditional Love   114

Epilogue	116
Notes	118
About the Author	120
Recommended Further Reading	121

By the Same Author
Magical Places of Britain
(Wyldwood Publishing 2013) – A photographic travel guide to Britain's natural sacred sites with associated folklore.

Oh ancient one!
You who came down from the stars...
You who witnessed the shaping of our world...
You who have watched your children grow over long ages of time...
You who stand among the gods!
Tell me now, of when the world was young...

Introduction

There was a time when we lived completely immersed in nature, from the wind in our hair to the earth beneath our feet, in constant contact with natural forces, interacting with them in every moment of our lives. The primal mind perceived these natural forces as sentient beings infused with spirit, and these spirits were immanent within all nature. This was our reality in those times, it was the world that we lived in, and it came with a sense of belonging and a feeling of connection to all things in the world around us.

Today, modern Westernised society has now become more separated from nature than any other culture in the history of the human race. We spend our time living in artificial environments, staring at flat screens and purchasing pre-packaged food that has been mass-produced on factory farms. Meanwhile rampant capitalism and globalisation are rapidly eating up the entire planet, consuming its resources and spewing out pollution into all corners of the Earth.

So how did we get here? How did we become so separated from the natural world that gave birth to us? How did we lose our intimate connection to the Earth and to the natural forces that once surrounded us? Why do we now destroy our own planet, the very earth we stand upon? How did we become so lost?

To answer these questions we will take a journey back in time, back to our primal origins when we and the land were one. Back to the time when there was no separation, to the time when we had primal awareness, completely immersed in nature and in harmony with the rhythms of the Earth.

During our journey we will investigate, at each stage of our development, what occurred that caused us to become separated from our natural state of being, and we will discover ways to turn back the wheels of time and reconnect with our primal origins.

As the veils of deception that have blinded us for centuries are slowly lifted, a new awareness will begin to shape itself. A feeling of hope will be rekindled for our future, for our planet and for the whole human race. A return to our true source and the joy of feeling connected with the Earth once more.

We will start our journey at the very beginning, back when we experienced a total immersion in nature and a complete connection to the Earth. We will attempt to enter into the primal mind and experience the world through their senses.

To enter into the primal mind we must first enter into the world of spirits. So let us journey back there, back through long ages of time, carried on the deep resonant voice of one of our ancient wisdom-keepers. A calm, steady voice that carries our deepest ancestral knowledge...

Part I

The World of Spirits

Long ago... Long before my ancestors raised stones or planted corn... There was a world of spirits...

There were spirits in the air, spirits in the water, spirits in the earth, and spirits in the mountains.

In the time before time, great primal spirits bent and shaped the Earth.

Spirits of air assaulted the land with fierce winds, brewed up tumultuous storms of thunder and lightning.

Spirits of water formed raging rivers; carved out gaping canyons; filled immense lakes.

Earth spirits shook and shaped the land; threw up mighty mountain ranges; moved whole continents.

Spirits of fire surged deep beneath the earth; erupted from the mountain-tops in great plumes of fiery magma; formed rivers of molten rock.

All the while the Earth basked in the radiance of the all-powerful Sun who bathed the world each day in his light and radiance. The Moon too cast her glow in the darkness, while the star spirits twinkled in the great dome of the cosmos.

These mighty forces each played their part in the shaping this world, as the Earth turned, the tides rose and fell, and the seasons passed endlessly by.

Life covered the Earth in an intricately complex weaving, filling every crest and crevice, every wide expanse and dark depth of this untamed primeval land. A kaleidoscope of shimmering creatures filled the oceans and rivers, tramped through the forests and the marshes, and swept across the open plains in great herds. They leapt high into the mountain crags, darted through the skies and crawled in the deepest caverns, forming an intricate web of life subject only to the mighty and mysterious forces of nature.

These forces caused the plants to grow and the waters to flow; they carried leaves on the wind, and shook boulders from the high mountains. From the tiniest insect, flower or pebble, to the great elemental spirits of earth, air, fire and water, everything was *governed by spirit, for everything was spirit.*

There were spirits of trees, spirits of leaves and roots and branches, spirits of the forest and spirits of the land – worlds within worlds within worlds – a great hierarchy of spirits connecting the vast cosmos to the intricate workings of life. Everything was a part of this spirit world, and everything was a part of something greater.

The leaf was a part of the branch, the branch was a part of the tree, the tree was a part of the forest, and the forest itself was just one small part of the great tapestry of life that enveloped the whole living Earth; a limitless expanse of thriving and vibrant wilderness that stretched from shore to shore, and from ocean depth to mountain height.

From this mysterious web of life emerged our most ancient ancestors, the first humans. It was a time when the people and the land were one: there was no separation. The Earth was a living being and we were a living part of it, at one with the greater flow of life.

Fully immersed in nature we could sense with a subtlety that would seem almost supernatural to us today; we were alert to every minute change in the air or soundscape; we instinctively knew where to travel to find food; we were guided by our senses alone to survive and to reveal to us our reality, without any conscious thought processes being necessary. Like other living beings we were tuned in to the changing of the seasons, the stations of the Sun, and the phases of the Moon.

This was a time when nature provided and spirit guided; a time of oneness; a time of connection with all Being. For the spirits were all around us, always – a breath of wind, a flight of birds, the sighing of a bough or a clap of thunder – all could signal the passing of spirits, for all was spirit.

The spirits provided and the spirits guided, our entire world was conceived of in this way, but the spirits were not separate from us, not something out there confronting us. For we were a part of that spirit world too, we had only to flow as part of that world, as we always had, and we would endure; the Earth would satisfy all our *needs.*

So this was the spirit world of old, the mysterious age of our primal ancestors, the mythical Dreamtime when we walked the Earth as one with the greater flow of life; a time when we relied on our heightened senses alone to survive and to reveal to us our reality. It was a time of oneness, of connection with nature and with all being in the ever-present realm of spirit.

The shaman's voice reverberated throughout the cave as he spoke to me of those most ancient of times, a time when mankind and the Earth were one. As I listened to his haunting words he revealed to me a world of spirits that existed in the depths of time before the mind of mankind had even begun to transform into its present state.

His powerful words were followed by a heavy silence, disturbed only by the occasional crackling of wood as the gently glowing fire illuminated his deeply lined face. He looked weary, drained from the exertion of channelling such ancient ancestral memories.

I waited in silence as he slowly opened his eyes and gazed at me. I gazed back at him, trying to take in all that he had just told me, eager to understand and to know more.

'Tell me, what happened to the world of spirits? Where did it go?'

His gaze intensified as his eyes seemed to penetrate the deepest recesses of my soul. Then slowly he opened his mouth to speak as his eyes darted around the walls of the cave.

The spirit world is here! It is all around us, always! And can still be experienced by those who have the senses to discern it.

'So why can we in the modern world no longer sense it?'

Because **you** have changed! You have lost the sense and the sensitivity for that realm through your many levels of separation from your true source. It is this ever-widening gap of separation that has been afflicting mankind throughout the ages. It has brought the human race to its present state, a state where they now pollute, pillage and destroy their own planet, the very Earth upon which they stand, their very source of life and sustenance!

The Earth has undergone massive changes. What was once a thriving and vibrant wilderness is now increasingly becoming a cultivated, developed and polluted landscape, sculpted to accommodate the wishes of only one of the many millions of diverse organisms that dwell here on planet Earth – mankind.

But it does not stop there... Mankind has now set itself upon a path to self-destruction – a path facilitated by the inexorable expansion of modern civilisation, with its rampant consumption, ignorant self-interest and increasingly destructive technological advances that are slowly eating up the entire planet and converting it into a soulless monoculture and polluted wasteland.

'Are the spirits still here then, among mankind's self-created desolation?'

The spirits of nature have fled! They have fled from the cities, from the highways, the industry, the pollution, and the man-made constructions that now cover the land. But always they are trying to return, seeking a way back in – carried on a current of air or through a crack in the pavement, or to a wind-blown corner of wasteland, where they can establish a foothold, before emerging once more to reclaim this land. For the spirit world is timeless, its patience infinite, its influence unrelenting!

'What then is this spirit world of which you speak?'

Now that is something that cannot be explained to you here in words! It would be like trying to describe the flow of a dance; it is something formless, intangible, which can only be recognised by its physical manifestations in this world and not by its true essence.

Take away the physicality of the dance and what are you left with? How do you define it? How can you define the flow of a river without the water, the growth of a tree without the wood, a breath of wind without the air? These things are formless, intangible, and yet they give form and make form possible in all its manifestations.

This formless and intangible world, the world of spirit, formed the reality that our ancestors lived in. But as our perceptions have changed so too has our very concept of reality changed. We live now in a world restricted by our own thoughts and beliefs, our preconceptions and our labelling. We are no longer free; we have become prisoners of our own thoughts and self-projected realities.

'How can I escape this mental prison?'

All you need to do to change your reality is to change your perceptions. But this is a task not so easily achieved, as your perceptions of reality are so deeply ingrained in your very being, in your culture, your society and your upbringing. First roll back the layers of separation that have parted you from the true essence of who you really are. Become one again with the source of all your being, with the very spirit of the Earth. And act soon, people must awaken before it is too late, and the world is despoiled beyond all repair, and no longer a fit place to live in for neither man nor beast.

'So how can I bring myself back into alignment with the spirit of the Earth?'

Tap back into the ancient wisdom, the source of all knowledge, and integrate this knowledge back into the modern world. By doing this will you be able to change your perceptions, and thereby change your reality. Combine ancient wisdom with modern learning to create a whole new paradigm for the future.

'So how do we remove these layers of separation and become connected with nature again, with the spirit of the Earth?'

To understand that you must first understand how you became so separated, and that, my friend, is a long, long tale...

Part II
Nineteen Levels of Separation

1

The Naming of the Spirits

As the minds of our early ancestors developed, so we started to become creatures who were living in two worlds, the world of spirit and the world of the mind. The mythical dreamtime was coming to an end as our reality started to be shaped by our perceptions. The rational part of the mind wanted to make sense of the spirit world, and so this process began with the naming of the spirits.

Every manifestation of nature was seen as having a spirit which was its very essence. Rocks and crystals, plants and animals, mountains and rivers – each had a spiritual aspect and these spirits were named and visualised as spiritual beings by our early ancestors. Even the forces that set our world into motion – the gusting wind, the flowing waters, the heated earth and the icy frost – also came to be seen as separate sentient spirits.

No longer were we just seamlessly interacting with these mysterious forces that shaped our world, they came instead to be viewed as spiritual beings with name and form and personality. Thus by naming them, and visualising them as living entities that we could interact with and influence, so we started to become separate

from them and see them as something 'other', something that was outside of ourselves, rather than something that we were also a part of.

These words and labels were our first level of separation from nature. As the spirits became labelled and personalised in this way, so each started to be viewed as having its own unique personality and characteristics.

The domain presided over by powerful spirits was also seen to be filled by lesser spirits, so as there was a spirit of the sea, a spirit of the forest, and a spirit of the mountain, so there were also spirits of the waves, spirits of trees and spirits of rocks. Water spirits swam in the waves, tree spirits haunted the forests and earth spirits dwelt beneath our feet.

So the naming of each of these spirits – sylphs in the air, undines in the water, dryads in the forest, or gnomes in the earth for example – now projected onto each of them a personality and a form which they did not previously possess, a symbolic representation of their true essence. These forms and personalities were a projection of our developing human consciousness; they were being forged in the minds of human beings. These forces of nature were very real, but our concept of them came from our own projected reality and thought forms.

The truth had now been distorted by the very act of naming it and defining it. The mysteries of the spirit realm could not be so easily defined. The mind of mankind has a need to simplify, to classify, and to label, in order to understand and to comprehend, but the mind is limited by its own words and thought forms, while the world of spirit that is manifest in nature is limitless and free of any such restrictions, and is able to flow seamlessly with the world around it, unhindered by any such classifications or labelling.

I ask you, where really does one ocean end and another begin? When does a stream become a river? How many trees must stand together to make a forest? All such manmade classifications are ultimately futile for all of nature is one; it is one continuous flow, a labyrinth of interweaving and interchanging energies. The decrease of one flows seamlessly into the increase of another, there is no separation.

Look at an oak tree and what do you see? It is not simply a 'tree', this convenient label we give it to facilitate communication, it is a wonder of creation, a whole living eco-system. There are things living on the tree and in the tree, there is bark and branch and leaf, there is a whole hidden world living underground, and there are the microbes and bacteria that give it life. Leaves rot into the earth, and the earth passes nutrients back through roots to the tree. Water is drawn up from the ground and as the leaves transpire it is passed into the air. Carbon is absorbed from the air to build mighty boughs, and sunlight produces oxygen from the leaves that we might breathe the breath of life.

Tree is just a convenient label we use, something to separate it from its environment and from us, an artificial label which we link to a simple word. By calling it a tree we diminish it, make it ordinary, dismiss it and take it for granted, we no longer appreciate the full wonder of its timeless creation.

In nature there is no separation, everything is connected; everything is a part of something greater. Everything is dependent upon the whole, and the whole is likewise dependent upon it, as a living breathing part of the great web of life. From an acorn, to a seedling, to a mighty oak of root and branch, to a rotting pile of compost, what truly is a tree? There is a unique undefinable essence that gives it life and form, but that essence itself is a part of something far greater.

Today we think of the air as nothing more than a gas, but to the primal mind it was a mysterious force infused with spirit. We speak of winds, breezes and gusts, but to the primal mind these were the spirits at work, the mysterious forces of nature made manifest. The spirits of air connected all things upon the surface of the Earth, from our breath to the treetops to the birds in the sky. Through our breath and our voice we had a direct connection to this ethereal world of spirits, and through chanting and prayer we sought to commune with and influence this spirit world.

Words had their uses, they facilitated communication and exchange of knowledge, but true knowledge and true understanding can be achieved only in silence. When the words are quieted, only then can the real essence of something be felt and revealed.

This separation created by words is why silent contemplation is so important, it is a way to consciously shut down the mind and the word horde, and allow us to return to that purest form of connection, the simple connection we have with the Earth and with spirit.

This sense of connection can also be achieved by other means; activities which bypass the mind and connect us directly back to our source. Simple pursuits such as fishing, handcrafting, dancing or walking in nature will take your awareness outside of yourself if you can but take the time to connect. Whenever you silently tune into your surroundings you are entering the world of spirits. Just as the repetitive steps of a lonely traveller induce a hypnotic state that can raise an awareness of the elements all around – of the air, and the earth, and the chatter of the wildlife – so the fisherman can silently connect with the spirit of the water, the wood carver connects with the spirit of the wood, and the dancer connects with the spirit of the dance.

Yes, words were our very first level of separation from our true nature, words and language and the thought processes associated with them. In Aboriginal culture the *miwi* (or soul) is depicted in a humanoid form but always without a mouth. The soul has no words.

Can you stop for a moment and imagine a world without words? Can you imagine a thought without words? Without words there can be no analysis, only acceptance.

So try shutting down the endless chatter of the mind for a

while, connect directly with nature and the world around you, experience oneness, and experience the connection that comes from silence.

And so the shaman instructed me to close my eyes and focus instead upon my breathing. A long deep breath in through my nose, which I held for a few seconds before breathing out through my mouth, long and slow, until my lungs were completely empty. I sat in the silence and emptiness for a couple of seconds more before my body naturally started to inhale again, and so the process continued.

He taught me how to breathe like an animal, using only my belly, just as we all did when we were still innocent babes, inflating my belly as I breathed in, and deflating it as I breathed out, without expanding or contracting my rib cage. He explained to me how only humans, and creatures under stress, breathe with their chests, and how in this stressed state it is not possible to let go of the mind that imprisons us, and to merge with our surroundings.

As I emptied my lungs again the shaman now instructed me to empty my mind too. Any thoughts that came into my head simply drifted away on my breath. I was instructed not to pay any attention to these thoughts, not to give them any energy, but to simply acknowledge them and watch them drift away.

As I breathed long and deep for ten minutes or more I felt my mind starting to clear until all there was my breath. And still I continued the breathing, long and slow, until my whole body became relaxed and felt heavy. It was now almost as if my body was sleeping, and my mind was sleeping too, but my consciousness was wide awake and free to drift outside of myself for a while. The mind and the body were left behind; I was just consciousness and awareness without any thoughts or judgement.

In this state of complete relaxation and silence I was now told to take my awareness out into the cave. I could sense the fire and its crackling flames, and I could feel its presence. I could sense the cold hard walls of rock, and noticed how different they felt to the warm glowing fire. I could also feel the fresh air gently drifting in through the cave entrance, which again had a very different quality. All of this happened without words, without judgement or analysis, simply acceptance and acknowledgement. I could *feel* the fire. I could *feel* the rocks. I could *feel* the air. I could sense the unique undefinable qualities of each. I could sense its 'spirit'!

So now I understood. I could feel the spirit in each thing, but there is no way to adequately define these feelings or put them into words. I can say the words **fire**, or **cave**, or **air**, but I understand now that they are just words, and that each will be interpreted differently by each person who hears these words, based upon their own past experiences and pre-conceptions about these things. Words can only ever be an approximation, for the underlying truth of something, its true essence, lies much deeper, and can only be experienced in silence.

So language has placed an artificial barrier between us and the truth. Language and all the mental constructs associated with it are ultimately at the root of all our separation. This process started with words and explanations, but soon enough these words evolved into whole stories and mythologies...

2

Stories & Myths

As our minds developed and our word horde grew, so we felt the need to account for all kinds of natural phenomena. Out of this need stories emerged, stories of spirits, stories of animals and stories of ancestors.

In some cultures rocks hurled down mountains were said to have been cast by primeval giants or great ancestral spirits; the flow of a river was created and controlled by the actions of a river goddess; and canyons were carved out by the passing of a mighty primeval serpent.

The whole sacred landscape of the Earth was said to be created by these great ancestor spirits; but these 'ancestors' are not our deceased forebears as we would think of them, they are the natural forces that form this land and all the beings that exist here on Earth.

The Earth in its primal state was moulded by these great forces that worked to produce the rocks and landforms, the plants and animals, and eventually us, the people.

In Aboriginal stories told about the dreamtime these ancestor spirits take on many different forms, alternating between people and animals and landforms. These stories served many functions in past times, they first of all familiarised people with the land and introduced them to its many facets and features. The minds of these pre-literate people worked very differently to ours, they did not digest facts, figures and lists like us, but transmitted their knowledge orally in the form of songs and stories and metaphor. These stories of the dreamtime ancestors traversing the land are known as *songlines* and they explain how the landscape is formed.

These ancestral songs can be seen to represent great 'vibrations' of energy, and it was the great vibrating 'bodies' of these ancestral spirits that formed the songlines and created the land.

The songline stories were also a survival mechanism, allowing people to remember natural features of the land and so travel safely from one place to another. The songline stories also taught Aboriginal *lore*, which is a universal law that describes a way of living in complete harmony with nature that is sustainable and enduring.

Timeless tales were woven in and out of the great cycles of life, the Sun and the Moon, the seasons and the stars.

So now everything in the world came to be accounted for by these stories, and our lives too were woven into these great narratives as we accepted our role in the great story of life.

Words and stories were a unique creation of the human mind. Animals have no stories, they do not analyse or rationalise, they act only from instinct and intuition, and vocalise only in the form of calls and signals by which they keep in contact or inform one another.

Our words had created an alternate reality for us, a reality where first everything was labelled with words and then explained with stories.

'So are the indigenous stories just fabricated tales with no basis in reality?'

When you read these ancient indigenous tales do not take

them literally, do not seek to place them in an historical or literal context; instead look for the hidden meanings behind the words. Imagine how these stories evolved out of observing nature and the world of spirit, and open yourself up to the mysteries that are being revealed. Don't get drawn into the literal interpretation of stories, mythologies and scriptures. The ancient wisdom contained in these stories can provide a pathway back to the world of spirit and a more intimate connection with nature. So after hearing a story, contemplate it in silence, turn off the chatter of the mind and enter a state of deep meditation using the techniques I taught you earlier, and try to gain an insight into the hidden wisdom that is being revealed by these tales.

Acknowledge that the spirit world is timeless and manifests itself in mysterious ways. Open yourself up to an understanding of the timeless principles that are being expressed in these tales.

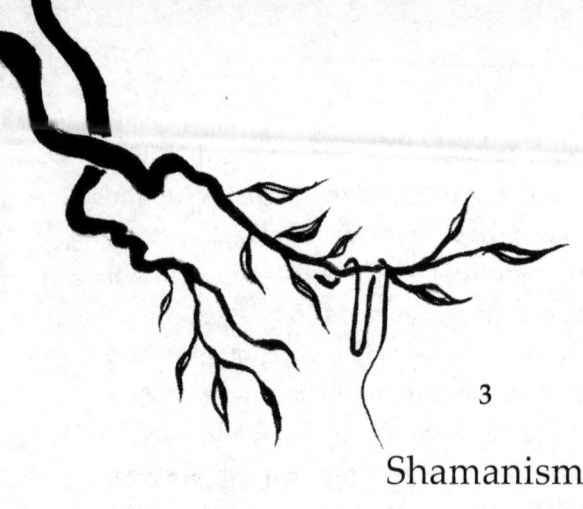

3

Shamanism

We were all shamans once, and we once all travelled freely within the world of spirits, but as our minds and our language expanded so our separation from the natural world increased, and we started to live instead in our stories.

Stories were repeated around the campfire as people sought to explain the phenomena they had experienced in the natural world in terms that the ever-inquisitive mind could more readily accept. Stories were invented to account for all kinds of phenomena and it was not long before people started to interpret these tales in new ways that were not originally intended.

But there were also those who remained more connected to the deeper mysteries of life. These were the wisdom-keepers and the shamans, those who understood that the stories were never meant to be taken literally, that they were tales designed to give the mind access to the workings of the mighty and mysterious forces of nature, the hidden knowledge of myth and metaphor that could be passed down through the generations.

But the ignorant, those who didn't understand this, took the tales literally, really believed in them and even embellished them, and repeated the stories without any idea of their true meaning. So over time the stories drifted further and further from the truth, until the original knowledge could barely be discerned. Stories became culture, and culture became beliefs and self-identity, until only the wise elders and the shamans still understood the true meaning behind the tales.

And thus was born the next level of separation, there were those who knew: the shamans and the wisdom-keepers, and there were those who were ignorant, who no longer understood their true connection to the realm of spirit. As time passed some of these shamans became priests and they too began to believe their own tales and to create their own stories for others to follow, but that we will save for later...

The shamans developed techniques to bypass the mind, and the world of stories that had started to dominate our thoughts. These techniques allowed them to journey freely again into the world of spirit. Some shamans consulted plant spirit helpers through the use of psychoactive drugs, while others entered altered states using repetitive music, dancing or chanting. Once in a trance not only could the shamans sense the spirits all around, they could also travel freely through the spirit world to gain vital information, about food sources, hunting, weather or disease.

In some groups everyone was a shaman and engaged in these activities as a matter of routine, but in other groups it was left to those with special abilities, and so the less gifted people came to rely on these shamans.

The separation created by the human mind and its stories had therefore been further compounded by passing the responsibility for contacting the spirit world on to others who specialised in this. The common people had their stories about the gods and the spirits which they told each other around the camp fire, but they consulted the spirits or the shamans only when they needed them. Offerings would be made or rituals would be performed to ensure a good hunt, or to bring rain and fertility. People still had interaction with the spirits and could even sense them, but their worldview was changing, a worldview that came to be dominated by their stories and a passing of responsibility for contacting the spirits over to those who had more authority in that field.

'Can I journey into the spirit world and contact the spirits in this way, using psychoactive drugs for instance?'

Nowadays many people take psychoactive drugs with no idea of what they are doing; they do it for fun or for recreation, finding it amusing, or sometimes terrifying. They are so far removed from their natural source that they cannot comprehend what they are experiencing, although to some the experience permanently expands their consciousness to a new level of awareness, or at least brings forth an appreciation of it. But these drugs can also cloud the mind, and have unforeseen side effects if not taken as part of a sacred ritual with an experienced shaman. It's better to contact the spirits in a more gentle and natural way, through deep breathing, meditation and sinking into stillness, immersing yourself completely in nature for extended periods of time. This will prepare your mind and your spirit for a deeper connection, one that will offer greater clarity in the long run for one who is just embarking upon this path.

Once you feel a sense of connection with your environment you can start to journey into the spirit world that surrounds you, taking your awareness outside of yourself for a while and into the spiritual landscape.

There are many techniques for doing this, for journeying outside your body and gaining an awareness of the many worlds that overlay this one. But I will show you how to journey using the drum.

And so the shaman produced a large skin bag and out of it

brought forth his drum. It was round and flat, about two feet across, and consisted animal skin stretched tightly over a wooden frame. He also produced the beater, a rib of elk with a padded tip.

Start by deep breathing and sinking into stillness, like I have already taught you, and then listen to the rhythmic beats of the drum.

He started to beat the drum in a fast but steady beat, the deep booming sounds reverberating throughout the cave. Very soon the hypnotic sound started to induce a trance-like state that absorbed my awareness.

Turn off the chatter of the mind and focus instead upon the resonance of the beats. Feel your awareness being consumed by the rhythmic sound of the drum and let impressions come to you. These may be stimulated by your environment or your recent experiences, or they may emerge from deep within your subconscious. You may be carried into a forest, or to a distant seashore, a hillside, a cave or a sacred site. Let these impressions develop. It may feel like daydreaming at first, but images, impressions, sounds and feelings will eventually form and rise to the surface, creating dreamlike experiences known as *journeys*. You are now journeying into the realm of spirit. This is the realm of your own subconscious and the 'dreaming' of the land around you. These stories and impressions are teachings, revealing some of the hidden mysteries of your soul's journey and of the spiritual landscape. You have opened yourself up and allowed yourself to be infused with spirit. You have received *inspiration*!

4

Altering our Environment

Our most ancient ancestors lived in small roving bands of extended family groups. They rarely saw other people, and when they did they would simply exchange news and information or, more rarely, they might settle disagreements, or arrange marriage partners.

The world at this time was one vast, endless expanse of wilderness for us to roam in; nature in the raw surrounded us and immersed us, and pervaded every aspect of our lives. Mankind was part of this endless flow of nature; food was simply hunted or gathered from the land. Mankind had its role to play in the huge web of life, lived in harmony with the nature spirits, was in communion with the plants and animals, and so acknowledged and harvested each in its own appropriate way which was in alignment with the greater flow of ever sustaining life on Earth.

A mutual respect and understanding, a symbiosis, existed between predator and prey, between hunter and hunted. Because they were all part of the greater flow of life, the vast interconnected web that encompassed all beings on this planet Earth, each fulfilling its own role in a harmonious, timeless and endlessly

sustainable way. Predators were kept fit and healthy by their need to catch prey, while the prey species were kept healthy by having their sick and weak members selectively removed by the predators. This check on the population of the prey species by the predators likewise allowed plants to grow and prevented overgrazing.

But change did happen, natural catastrophes could sometimes occur and species did become extinct; but always the harmony would return, and a new balance of coexistence would settle upon the face of the Earth. Individual species may suffer temporarily, but Mother Earth maintained her vast and nurturing body in harmony. Over millions of years, great eons of time, she has developed ways to absorb every kind of catastrophe.

As human beings spread over the face of the Earth, the mind of mankind developed, and so ever more ingenious ways were employed to acquire our food. In some cultures hunting weapons and fish traps were developed, also corralling of animals to capture them or kill them in pits or deadfalls, and clearing of the land by burning or felling to promote new growth and to facilitate hunting. Each of these methods could be employed to make the life of mankind that little bit easier, and to make abundance more assured. Mankind was no longer simply living from the natural environment, guided by the spirits of nature, but instead tried to manipulate the environment in subtle ways, and used ingenuity in order to increase abundance.

Spirits were contacted during this process – the spirits of the land, the spirits of the hunt, and the spirits of the hunted – in order to ensure that the time and place were right, that the omens were favourable, and that abundance could be maintained.

Some people began to plant seeds, knowing that if they came back to this spot later in the year there would be food there to harvest. Others started to domesticate animals, and to provide them with food or pasture. No longer solely reliant upon the wild spirits of the hunt, people instead started to commune with the spirits of the land and the weather to grow their crops, or to protect their stock. The world of spirits was still very real to them, but now they were attempting to control their environment in subtle ways in order to survive, they no longer simply trusted in the ebb and flow of life to provide for all their needs, they now wanted an insurance policy: a store of grain or a herd of livestock that would see them through the hard times. And so they prayed to these fertility spirits, and left them offerings, and lived in fear of their displeasure.

This process of altering the environment to suit our needs is still going on today. Some countries are now so altered that practically no natural environment still exists, all has been turned over to farming, grazing, industrialisation and urbanisation, while in other countries the areas of remaining wilderness are being constantly eaten away and infiltrated by roads, logging, mining and development.

'So where now can we find the wilderness which will heal our souls?'

Finding true wilderness in our world is becoming increasingly difficult. You may have to travel to a National Park, a remote mountain top, or a distant sea shore. Once there become immersed in the sights, sounds, and smells of the place. Make camp and forage for food if you can. Wandering the wild open spaces we can interact directly with the beings that dwell there and with the spirits of nature; the plants, animals and rocks, and the elements of earth, air, fire and water; and become lost in the wonder and majesty of it all.

'How can I experience this intimate connection to nature and to the land once more?'

First of all find a remote and secluded place in nature where you will not be disturbed, preferably a place by water. A clearing by a woodland stream on a warm sunny day would be ideal. Leave behind all the trappings of civilisation, all

your clothes and jewellery must be removed, everything! And then walk naked into the space in a pure and primal state, carrying nothing more than nature provided for you at birth. If you are wearing perfume, deodorant or make-up then wash it all off, and scrub your skin with soil taken from the earth around you. You do not want to be carrying any scents of civilisation with you. Rub the dirt into every part of your body until you smell like the earth yourself.

And then, silence... Keep silently still... Breathe gently... Observe and absorb...

Take in the sights, sounds, smells and sensations, all that you can register with your senses; the sunlight illuminating the leaves, the sighing of the breeze or the rippling of the water, the smell of the earth or the scent of flowers, the gentle caress of the air on your exposed skin.

Now, gently, slowly take a few silent steps. Notice with each step the feel of the earth and rocks beneath your feet. You are a part of the forest now, a part of this landscape. Notice how vibrantly alive everything is, how sentient. Slowly and gently start to explore the area around you, as if you were a curious primal being, seeing and experiencing things for the first time... Wade through the water and feel its coolness and fluid texture. Climb a tree and feel the bark beneath your toes. Feel the freedom and sense of connection experienced by our primal ancestors. Now lie down on the grass or the earth and feel the texture of it against your body as the earth supports you. Look up at the sky and observe the clouds silently sailing by.

This is how it feels to be in love with Mother Earth! To be in contact with her in every moment, to be immersed in the sights, sounds, smells and sensations of her. With arms outstretched, feel the warming rays of the Sun gently caress your body.

Now you have connected with all four of the elements; the

earth beneath you, the air around you, the flowing water and the fire of the Sun. It is now time to connect with the spirits of this place. Send your awareness out and *feel* the passive presence of the earth spirits in the rocks and the soil. *Feel* the spirits of the air as they swirl and pass by you in soft gusts, gently agitating the leaves and the grasses. *Feel* the spirits of the water as they gurgle and splash, washing over rocks or swirling in eddies. *Feel* the spirits of fire beaming down from Grandfather Sun on rays of flickering light, warming the earth, the air, the leaves and the boughs.

Notice how everything is interacting; dancing and playing in an endless game; the air, the water, the leaves, the earth and the sunlight dance around you and interact with one another, each in their own unique way. The spirits are coming alive for you now, you can feel them!

Now feel the sensations in your own body too. Feel how it is interacting with all the other spirits around you, disturbing the passage of air, absorbing the heat of the Sun, rooted firmly into the earth that supports your weight. Splash some water on your body, rub leaves onto your skin and roll on the ground.

Can you feel the spirits of nature now? Feel them all around you, interacting, flowing, merging, each with its own unique quality yet each an integral part of the whole.

Feel how the trees are slowly growing, reaching up to the sunlight. *Feel* how the insects are navigating, carried on a current of air, each on its own unique and mysterious mission. Everything is in motion and constantly changing, each at its own pace. The Earth is slowly turning, the trees are slowly reaching out and growing, the grass is gently swaying, the water is ceaselessly flowing, the air is rushing by, the birds are calling, and the insects are buzzing around you in a merry dance. They belong here, as do you, naked and covered in the earth of this land. You are a part of this world!

Feel how vibrantly alive everything is, how sentient, how infused with spirit force! Feel how vibrantly alive you are too! This is how it feels to have primal awareness! Experience it fully, absorb it and remember it.

You could not have attained this awareness while in your office or driving your car, or while encased in modern clothes with rubber soles insulating your feet from the Earth. How then will the spirits of fire and the spirits of air warm and caress your skin, or the spirits of earth receive the imprint of your toes? All our stress, all our worries, and all our feelings of loneliness and disconnection stem from us not having this constant connection with beauty of the Earth. When we are making love to the Earth in every moment, we need nothing more.

So now is the time to dispense with the words and the mind games, and let go of the written word. Go out into the wild places and make a connection that feels truly *real*. Swim in the ocean, roll in the sand, embrace a tree, run naked through the forest and rejoice, for you are alive and reborn! No longer feeling numb, but instead feeling vibrantly alive, free, powerful and connected; a sovereign being and a meaningful part of the Earth!

There is no longer any need to shuffle around an office like you are one of the living dead, for you were not born a slave! Any servitude you imagine is just a mental construct, a self-created prison built from social pressures and obligations, imagined fears and comforting weaknesses. There is nothing more satisfying in life than to feel totally, completely and vibrantly alive! This can be achieved by feeling fully connected to your natural source, both physically and spiritually.

5

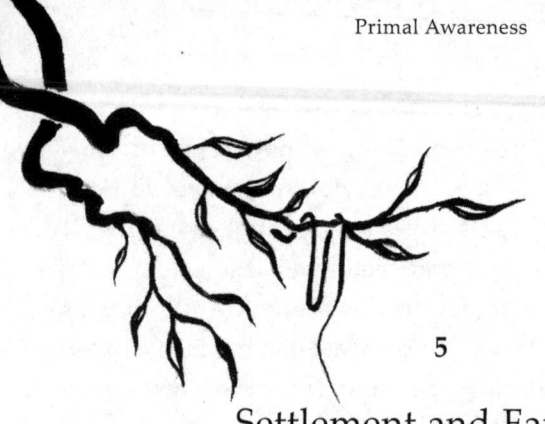

Settlement and Farming

While some continued to subsist by ever-more ingenious methods of hunting and gathering, others turned to agriculture and livestock as their main form of sustenance. Instead of constantly roaming to source their food, a regular food supply could now be kept close at hand. But there were consequences to be paid for this, the land now needed to be worked and people needed to become more settled, and so they experienced separation from their natural primal state. No longer were they fully immersed in the endless expanse of wild nature but instead became focused upon small plots of land.

As agriculture became more commonplace and more organised so more and more people began to settle down in one place and very soon the first permanent human settlements were born.

Tending to a plot of land required the building of more permanent structures to settle in. These manmade structures had to be imposed upon the natural environment. At this time mankind was still very much in contact with the world of spirits, they understood the flow of natural forces around them, and they understood the problems that building artificial structures would impose.

They knew that they had to be constructed in harmony with the surrounding environment, the local spirits of the land had to be consulted and propitiated, offerings had to be made to them, powerful places and spirit paths had to be avoided, and the siting of a building in a harmonious location was paramount. Once constructed it was assumed that local spirits would take up residence in the dwelling, and so shrines were constructed and offerings were made to these household spirits. Mankind had imposed upon the land, but mankind tried to make amends, so that harmonious existence would still be possible.

To the people working in the settlements the wild now seemed like a dangerous place full of malignant spirits and wild beasts. What had once been their natural environment, a place of harmonious coexistence, now became a threatening and dangerous place best avoided. So now there were two kinds of people, there were those in the settlements, in their cosy little dwellings, and there were those out there in the wilds, the wild nomadic people. An artificial barrier had been created between the settled farmers and the world outside, and they experienced more separation.

Interestingly the Bible story of Adam and Eve is an allegory of this fall from connection to separation. They were living naked in the Garden of Eden, they ate the fruit of the tree of knowledge, and have suffered the consequences of this knowledge and the pain of this separation ever since. It is telling that the sons of Adam and Eve were both farmers.

Nowadays the settled people have become completely dominant. Nomadic people, where they still exist, are shunned and mistrusted. The 'wild' is considered as something dangerous, and prolonged contact is avoided by most city dwellers. We feel safe and secure in our 'homes', but we also feel isolated and disconnected, something incredibly valuable is missing from our lives.

The very concept of having a fixed home was a direct development of this more settled way of life. Previous to this everywhere we roamed was our home. The very body of Mother Earth herself was our home because we felt connected to her and we belonged there.

'But didn't hunter-gatherers all die young? Wasn't theirs a brutal existence struggling against the elements? Surely people were better of when they started farming?'

The image we have of hunter-gatherers today is of pitiable half-starved people scraping an existence on the margins of society. This misconception has come about because long ago all the best land was taken from the hunter gathers by the more numerous and organised settled people who wanted the land for their farming practices. Therefore in places such as Southern Africa we are left with this image of Bushmen scraping a miserable existence in the hot desert, whereas in reality they would originally have occupied the best and most fertile lands where the food was abundant; but they were driven away from this land as the settled people took it for their farming.

Visiting a wildlife reserve in Africa today you can get a small taste of what this abundance would once have been like, teaming with herds of antelope, buffalo, wildebeest, giraffe, zebra, and all kinds of smaller animals, game would have been easy to come by. Some ocean shores and rivers were so abundant in fish and shellfish that permanent settlements could be supported there. Hunter-gatherers worked far less than their farming neighbours, and in general they were never short of food. Communicable diseases were almost unknown among hunter-gatherers, as with low population densities, diseases could not spread and so could not survive.

It is true that infant mortality was high among hunter-gatherers, weak and unhealthy children could not survive, and therefore life expectancy at birth was relatively low. But this has been misinterpreted by those who seek to prove that hunter-gathers died young and never lived to a ripe old age.

However, hunter-gathers, once they survived infancy, were

generally fit and healthy, and had a perfect diet that they had adapted to over tens of thousands of years, a diet from the land out of which they had evolved. Tribal elders and wisdom-keepers were an integral feature of indigenous cultures; survival to old age was in fact commonplace.

Farming put an end to this healthy way of life, soon people worked too hard, they ate the wrong food, and were subject to all kinds of diseases and infestations. These developed through living in close contact with one another and with their animals. Archaeological remains show that settled farming people became shorter and weaker than the tall and healthy hunter-gatherers; essentially they were malnourished. These settled communities soon became unsanitary, and life expectancy plummeted.

The main advantages of farming and settlement were increased food surpluses and an increased birth rate, and the settled lifestyle meant that more children could be raised in one place, so despite the higher mortality rates the populations of farming communities were soon on the rise.

The people who built Stonehenge were Stone Age farmers, and likewise the people who built the Great Pyramids were Bronze Age farmers, their lives were hard and often short. All of this happened so long ago in prehistory that we simply assume that ill health and a low life expectancy is what all ancient humans had to contend with, but the life of a hunter-gatherer living in a rich and abundant environment was very much more healthy and satisfying than the one experienced by these early farming communities.

They lived a healthy lifestyle free from modern diseases like cancer and heart disease[1]; they lived in harmony with the environment, with a deep sense of connection to the earth.

'How can I experience this intimate feeling of connection with the land?'

You may benefit from spending time in the semi-natural

environment of a city park or an agricultural countryside, but only in true wilderness will you find the environment which will heal your soul.

Wander the wild places until you find a location which feels special to you, and then spend time there, observing all the beings that live there, the plants, the animals and the rocks, and all the beings that pass through there, the birds, the insects, the winds and the clouds. The more time you spend on the land the more you will feel like you belong there and are a part of it.

Find a peaceful spot and meditate, enter a deeply relaxed state, and the creatures there will feel at ease with you and become accustomed to you. Keep doing this until you feel completely immersed in your surroundings, until you feel completely connected and at one. Then gently walk barefoot upon the earth, sample some of the wild food that nature provides, and camp beneath the stars. Return at different times of the year; observe the change of the seasons, until you feel like you have become a part of this land.

'How can I be as healthy as a hunter gatherer?'

Staying healthy in today's toxic world is not easy. There is pollution in the air, pollution in the water, and pollution in the earth. There are poisonous chemicals, herbicides, pesticides, radiation, electro-magnetic interference, and genetic modification.

Many books have been written about the paleo-diet which mimics the diet of early hunter-gatherers, but unless you can avoid all the toxins and pollution it will avail you little. Try to make sure your food is fresh, unprocessed, organic, healthy and vibrant. Avoid processed, preserved, polluted and modified food. Ask yourself: was this food living just a few hours ago? If the answer is no then it's probably not something a hunter-gatherer would have eaten, it has already lost a great deal of its life force and vitality, and your body will be expending energy fighting off the toxins it contains. Too many of these toxins and over time you may develop cancer or some other degenerative disease. To remain healthy and whole you should eat food that is healthy and wholesome, not just occasionally, but as often as you can.

Exercise regularly, out in nature, and find ways to simplify your life and meditate often to quieten your mind. This will help you to reduce stress and maintain optimum health.

6

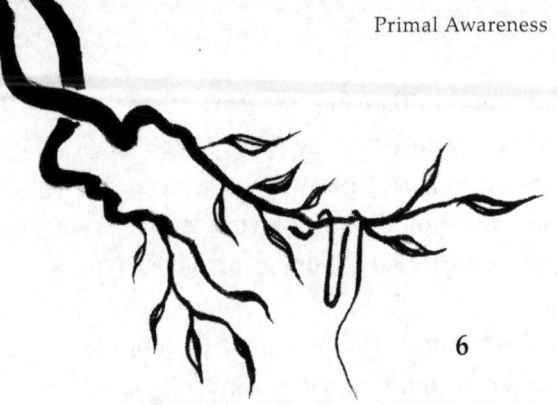

Tribal Culture and Division of Labour

The extra food generated by farming soon produced extra population, people started to live side by side, in a way that had been unnatural to them previously. Disputes sometimes arose in these more crowded settlements and so the need arose for some kind of administration. Previously all decisions had been made by group discussion until a mutual consensus was reached, but now chieftains were nominated in order to arbitrate disputes in these more complex societies.

The extra supplies also meant that now not everyone had to find or grow their own food, people could specialise and fulfil other roles in society. The advent of work, and specialised roles, signalled a new level of separation from our primal existence, especially for those who no longer worked the land.

Nowadays we consider tribal people to be the most primitive of people but this is far from the truth. Tribal structures were something new, highly organised and complex, they came with their own set of customs and taboos that soon developed into unique cultures as people came to differentiate themselves from neighbouring tribes. Tribes came with chiefs, priests, healers, specialised craftsmen and, soon enough, warriors.

Each had their own speciality, a job to do, which separated them from their natural role in the environment. People now became more focused upon their culture and their role in that culture, they had risen above the primal state of simply living from the land alongside nature.

We often hark back to the times when we were more connected to the land, to tribal times, without realising that even then the separation had already begun; the separation that is created by the human ego and the workings of the restless mind.

Tribes later amalgamated to become nations, or they had nationhood thrust upon them, and so cultural norms evolved into a national identity which came with its own new set of restrictions. The work that we do now separates us from the natural environment and the cycle of the seasons, and the culture that we live in has a set of unspoken rules that we are expected to conform to.

'So how can I break down the barriers that have been created by my cultural identity?'

Sometimes it can be difficult to acknowledge our own culture as there is so much that we take for granted and do not even notice. It may be more fruitful to observe another culture. What makes them different? What unusual customs do they have? How much of this is linked to cultural expectations that have been imposed upon them since birth? If we were all naked in nature and hunting our food, would we be so different?

Now look at your own culture, and your own beliefs and values. Why do we believe that our values are better than other peoples? Does it make us feel superior? More evolved? Acknowledge that most of what we believe and value is based mainly upon our cultural programming.

We are all just human beings doing our best to survive, given the work and the cultural restrictions that have been

imposed upon us. It is important to recognise this and to relinquish the 'them and us' mentality that keeps us separated. Connection comes from acceptance. So we should try to be accepting of all people and all cultures, no matter how strange their customs may seem to us.

Native cultures were once seen as abhorrent by Christian missionaries, who sought to impose their evangelising culture upon them. This is still going on extensively in the world today as wealthy Christian organisations send missionaries out to all remote corners of the world and seek to convert native peoples, severing their connection to the spirits and to the land that has sustained them for millennia, and seeking to impose their own belief systems upon them. The rich culture of native people, their centuries of tradition and their intimate connection to nature are still threatened by these actions.

We should not make the same mistake of thinking that we know better and telling others how to live. Acceptance is the key. It is important we stay in our power and set clear boundaries, so that others do not take advantage of us, but we should try practising acceptance at every opportunity.

7

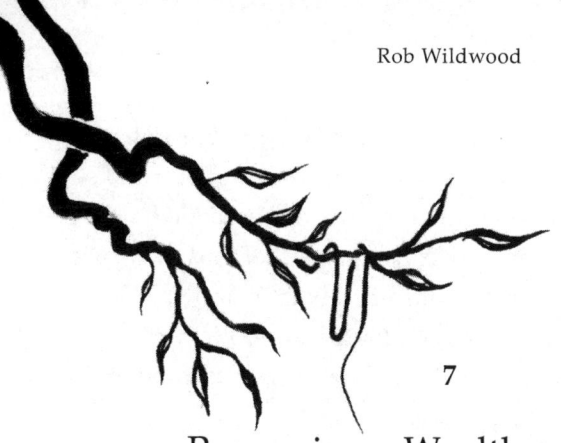

Possessions, Wealth and Status

More permanent dwellings brought an increased sense of security. People came to rely upon their dwellings and so developed the concept of having a permanent home. This home became filled with all kinds of objects that would not have been possible to transport in their previously nomadic lifestyles. Out of this evolved the idea of accumulating personal possessions. Our plot of land now belonged to us, the crops we had planted belonged to us, our farming tools belonged to us, whereas previously there had been no concept of ownership at all. This created another level of separation from the environment; no longer were we just a part of it, now parts of it could be owned by us or belong to us.

Previously all that we needed to survive could be readily obtained from the environment around us, even hunting weapons and more complex objects could be easily constructed, so ownership of an object made no sense because it had no intrinsic value and could be easily replaced using the materials at hand. Also, small extended-family groups of hunter-gatherers or early farmers would have shared all their food and would have distributed it communally. Hoarding was unheard of in these groups and would have been seen as contrary to natural 'lore'. But in larger tribal settlements this communal living was more difficult and so chieftains could emerge to mediate between the people and to distribute resources. In some societies these chieftains, realising that they had authority, became consumed with their own power, wealth and status, and so began the long road to hierarchy and domination. Once upon a time everyone was equal, but now some started to assume more power.

Certain objects were seen as especially useful or sacred, and these started to be traded over long distances. A certain 'value' was accorded to these objects, and people would be willing relinquish other things in order to acquire these valuable items.

Very soon it was noticed that some people had acquired more than others, had a better dwelling, more food or livestock, better tools. People started to get greedy, or envious. They came to focus more and more on their possessions and how much they had, and less on simply living in a harmonious way with their neighbours and with the environment.

Those with more possessions were able to gain a higher status within society and soon possessions began to be associated with self-worth and social standing.

Today the focus on possessions has become rampant, it is the main focus of many people's lives and is seen as the justification for all the unnecessary trials that we put ourselves through. To acquire a huge amount of possessions, be it money, land or luxuries, is seen as the pinnacle of achievement in our society; this is true in most societies in the world today and in historical times too, ever since the first human noticed that his neighbour had more cattle than he did. But there was a world before this, a world of sharing where possessions didn't matter.

'What motivated these people then? I cannot imagine it now.'

They were motivated by companionship and connection, good food and the freedom to roam the wild expanses of Mother Earth. When you are close to spirit and to nature then nothing else matters. To be in constant contact with Mother Earth, to be fully immersed in nature, is to have empathy with all beings and with the great mysteries of life; no material pleasure can compete with that! It is a constant feeling of connection, of being nurtured, of belonging. This is the feeling we are missing in the modern world, this is the empty gap in our souls that we seek to fill with 'things.'

There are still those on the planet who are able to live without any modern possessions. Their tiny numbers are dwindling daily but they have not yet been drawn into our moneyed economy. They live in deep jungles or on remote islands, and they support each other and share what food they are able to acquire.

'Could we really survive now, without money and possessions?'

It would take a great deal of courage and an intimate knowledge of the environment for us to survive for any length of time now with nothing more than our physical ability and our human ingenuity.

But we could start by trying to live more simply. Instead of focusing our attention on acquiring more status or more things, we can instead focus upon living in a harmonious way with our environment and with the people that surround us.

So let go of the constant struggle for wealth and status and instead embrace sharing. Share with each other and share the Earth with all the beings that inhabit her. Give back to the Earth as much as you take and trust that the Earth can satisfy all your needs, as it always has.

8

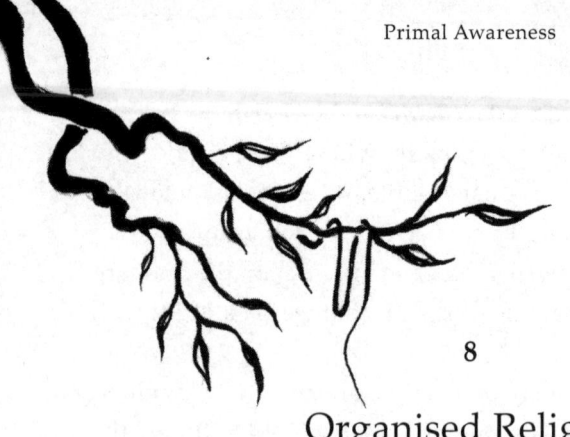

Organised Religion

In many cultures the spirits that were passive and receptive came to be viewed as female, while those that were active and seemed to project their power were viewed as having male characteristics. Thus was born the concept of the Earth being the mother and the sky being the father. The receptive Earth Goddess was viewed as female, as each year she was fertilised by the potent rays of the male Sun God, bringing forth new life each spring.

The Earth Goddess and the Sun God were soon joined in their pantheon by other powerful spirits such as the Moon Goddess, the sky gods, the gods of growth and fertility and many others, each with their own personality and unique characteristics.

So a spirit was viewed as the embodiment of a natural force or a natural phenomenon, and a god was simply a powerful spirit who represented one of the major creative forces in nature. Thus the spirit of the Sun, the Sun God, passed through the sky each day sending us his light and warmth, the spirit of the Earth, the Earth Goddess, sustained us and supported us on her vast nurturing body, and the spirits of fertility, the fertility gods, caused all things to grow and brought forth new life.

Priests fulfilled the role of mediating between these powerful gods and the people. They formalised the worship of these gods and developed all kinds of rituals and ceremonies to interact with them.

As the priests became more powerful within civilized society they began to organise themselves into a hierarchy. They developed a complex system of belief which only they fully understood. They claimed a direct connection and a special favour with the gods and spirits. The superstitious people came to subscribe to these beliefs and to revere the priests as intermediaries between themselves and the gods.

The priesthood also built temples to the gods so that they could be worshipped there. These temples were often constructed at powerful locations in the land to tap in to the energy of the Earth and the celestial bodies, and they were often built according to sacred principles so that they did not impose upon the land but rather were built in harmony with it linking Earth to sky. The priests were still attuned to the energies of the Earth and the energies of the cosmos, and they knew that all structures had to be built in harmony with the spiritual landscape, but they now sought to channel these energies into the fertility of the land and the life-force of their people, which led to the increase of their own personal power and status.

Flushed with this sense of power some of the priests started to use their power to manipulate and to control people, and thus they too became separated from the source of their spirituality.

As religions developed in complexity so their law books expanded and imposed all kinds of restrictions upon the people. Restrictions were placed upon certain foods, acceptable clothing, and sacred or taboo locations. Restrictions were also placed upon intimacy and human interactions, which became more formalised, so that the people were no longer at liberty to interact with one another in a free and natural way.

Through these restrictions, and the barriers placed between themselves and the spirit world by the priests, people became increasingly separated from their true nature, and were less able to express themselves in a natural manner. They had become subservient to the dogma of religion, a dogma created in the minds of the priesthood.

Our lives today are still restricted by these same religious beliefs, even if we consider ourselves non-religious. We are restricted by a society that developed from hundreds of years of religious repression and we would be interacting with each other very differently now were it not for all this social pressure that has been imposed on us.

'How do I break down these restrictive barriers that have been placed upon me by religion and society?'

Nowadays many people in the West are rejecting religion and seeking a new spirituality based upon personal growth and freedom of expression, each making their own personal connection to spirit in a way that feels appropriate to them. There are many ways to approach this new spirituality, and inspiration is being taken from all parts of the globe and is being integrated into a spirituality that is personal to each individual and not based upon any religious dogma or structured hierarchy. You are now free to connect with spirit in any way you choose. Churches, temples and priests are no longer necessary.

Inspiration can be taken from cultures all over the world, including the teachings of Celtic Druidry, Christian Gnosticism, Islamic Sufism, Tibetan Buddhism, Aboriginal lore, Chinese Taoism, Hindu spirituality, Norse Seidr, Siberian shamanism, Native American spirituality and a whole host of other indigenous traditions.

You could even go to the length of spending time amongst indigenous people such as Kalahari Bushmen or Amazonian tribes to gain an experience of total immersion in primal spirituality.

New methods of connecting with people on a spiritual

level and breaking down artificial social barriers are also being developed in the West, and incorporate self-expression and new ways of communicating and breaking down social barriers. You could try attending workshops in non-violent communication, Tantra, ecstatic dance and Biodanza for instance.

Carrying this new level of openness and acceptance out into wider society can be difficult though and will force you to notice how restricted many people still are by social norms. You may even experience anger or ridicule from people who believe that their way of behaving is the only 'proper' way to behave. Therefore, it is preferable to try to surround yourself with like-minded people who are on a similar path to spiritual reconnection. Don't attempt to 'convert' people to your way of thinking, simply remain true to what your inner wisdom is telling you and treat people compassionately. Then hopefully you will become an inspiration to others.

9

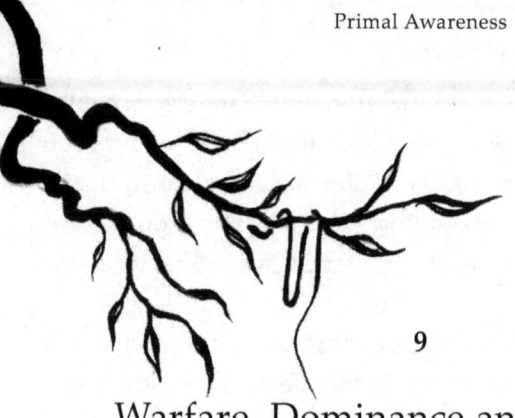

Warfare, Dominance and Patriarchy

As our resources grew so they now had to be protected from hungry and avaricious neighbours, and so a new class emerged, the warriors. Disputes with our increasingly populous neighbours now became more common, and in times of hardship we could decide to raid weaker tribes and steal their storehouse of food, and so the concept of organised warfare raised its ugly head, an unnatural act for the previously peaceful people to engage in.

In more 'primitive' societies warfare was ritualised and people rarely got killed; it was a show of strength and a way of restoring balance and harmony after transgressions had taken place. But in more organised and dominant societies a new form of warfare developed, one that was intent upon destruction, murder, robbery and enslavement. These aggressive tribes invented whole new mythologies and belief systems to justify their barbarous acts, following a patriarchal religion based upon male-dominated sky gods.

These bands of warriors and their chieftains soon came to completely dominate the land, as the peaceful Earth-based religions were swept away before them. These new beliefs created further separation from nature and from

Mother Earth, as a religion of peaceful co-existence came to be replaced by a religion of dominance. A belief in gods of war and a warrior's afterlife only justified more of this bloodshed and encouraged ever-more barbarous acts of murder.

Increased possessions came to be associated with status in this new patriarchal society. These included herds of cattle, stores of grain, and tracts of land, fine clothing, shiny weapons and personal adornments. Warfare became a means to attain these possessions and to improve status.

Warriors and chieftains became more focused upon themselves and their own standing within society, and so they became less focused upon their natural environment. This inner focus created more separation from nature as personal aggrandisement and tribal dominance became their main motivations.

Maintaining tribal identity and fighting for one's tribe came to be of paramount importance and served to keep people loyal to the controlling power structures. Later on, when nation states developed, patriotism and nationalism served this same function.

Peace can never be attained while greed and envy are given free rein. Peace is attained through acceptance. Accepting other cultures and beliefs, no matter how strange or misguided they may seem to us, and accepting that we have enough, and have no need to exploit or to impose upon our more vulnerable neighbours.

Even the most seemingly hostile and twisted belief systems are simply the product of people who are too wrapped up in their own cultural stories and separated from the true source of their being. On an individual level, all these people really need are love and acceptance.

'Can we really love our enemies and have compassion for those who have wronged us?'

Loving your enemies is not as impossible as it seems and can be achieved through developing empathy and recognising

that deep down your enemies are simply human souls just like you. While living here in this world we are governed by the forces of creation and destruction, but our inner spirit comes from a timeless and unchanging realm that is the source of us all. When looked at this way our enemies are simply our soul brothers and soul sisters who have been made manifest in flesh and have been brought up and indoctrinated into a different set of beliefs to ours.

This is true even if it is just your neighbour who thinks it's OK to steal from you, for instance. There are consequences to be paid for transgressing the law, be it local law or natural 'lore', but these people deserve our pity and compassion, not our hatred and bitterness, because given the right circumstances and a different upbringing we could be just like them. We are here to experience the joy of physical form and the teachings this world has to offer, not to deny life and happiness to others simply because we do not agree with their actions.

It is time for an end to war and suffering, an end to greed and exploitation, and an end to imposing our cultural beliefs onto others. We are all children of Mother Earth, born of her womb as brothers and sisters. We were born, not to compete, as modern evolutionary scientists and economists would have us believe, but to support each other, here on the living body of Mother Earth.

10

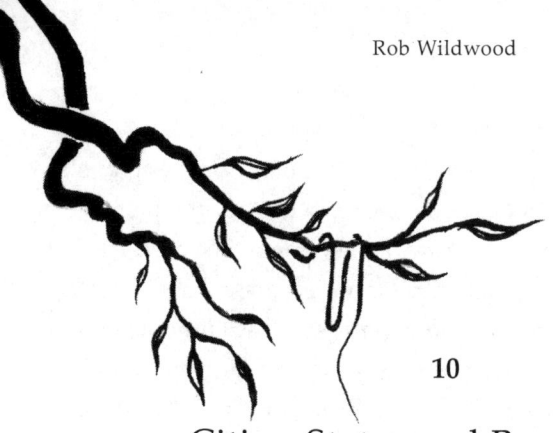

Cities, States and Bureaucracy

Farming increased our surplus food supply and enabled larger populations to grow. People who no longer needed to hunt or farm could take on specialist roles in government, administration or the priesthood. Specialists of all kinds became concentrated into densely populated settlements that became known as towns and cities. No longer needing to work on the land they were cut off from the spirits of nature, living in a world that was almost entirely manmade. They still worshipped at shrines and consulted oracles, but this was now the only connection they had to the realm of spirit. They had created an artificial world for themselves to live in, and because the ones with the power and control were city dwellers themselves, this artificial reality and its belief system were exported to the countryside too.

Religion had developed based upon the wishes and desires of the urban elites. Spirits were no longer consulted just to help sustain life and good health, now they were consulted to confer victory, wealth and power. What had once been 'power with', a way of living in harmony with nature and the spirit world, now became 'power over', a culture of domination, very much separated from its original spirituality.

Laws and regulations were imposed upon the people to ensure compliance with these beliefs, and to control the people's behaviour in these ever-more crowded settlements. These laws, which were necessary to maintain order in these large settlements, were then exported to the countryside too. Independent farmers who were once living freely upon the land were now subservient to the wishes of the centralised power in the cities, and had to pay them a share of their yields as a tax in order to maintain this central government. They saw a life of freedom reduced to one of servitude.

The rules imposed upon people's spiritual lives by the priesthood had now been joined by a whole new set of rules imposed from central governments in the cities, as all aspects of daily life became regulated and controlled.

To be a sovereign being and to be connected to the source of who we truly are is to have complete freedom to express ourselves. So as control and regulation grew, so freedom diminished and feelings of separation increased.

Much of our time today is taken up by the requirements of the bureaucracies that have been imposed upon us. We fill in forms, we pay tax, we visit offices, counters, banks and departments. We join queues and we waste untold time waiting on hold while computerised systems place us in artificial queues.

Human beings in the past had to contend with none of this! We were free, our time was our own, the world was our limitless playground. Is it any wonder that nowadays stress and stress-related illnesses are rife?

'How can I reduce all the stress that has been imposed upon my life?'

Take time away from the bureaucracy, and try to lead a simple, stress-free life for a while. What can you do to simplify your life? What can be discarded that is unnecessary? Must you sign that new contract? Must you purchase more goods on credit? Will they truly enrich your life and make you happy? All you really need to make your life happier is simplicity and simple pleasures such as nature provides for us.

Stop chasing after bigger, better, flashier. These things will not make you happy. Spend time out on the land, and take time to converse with people. These things are free and always have been. Try living within your means and living simply, feeling an intimate connection to the Earth and the beauty of nature that surrounds you.

11

Writing

As laws and trade developed so some cultures felt the need to represent things symbolically. Symbols developed into writing and soon abstract ideas could be represented through these characters.

Reading and writing signalled a further level of separation. No longer were humans just interacting directly with other humans, observing body language and sensing emotions, now they were interacting instead with a piece of parchment, or a wax or clay tablet.

This had the effect of creating a feedback loop between the reader and the parchment as they placed their own internal interpretations upon the written word. Previously this loop, this circuit of communication, existed between living people and communities, who could correct mistakes in communication instantly and put emphasis where it was needed, but now the words became fixed and sterile, devoid of emotion.

As stories came to be written down so they too became fixed. They became fixed not only upon parchment, but also in space and time too. In oral cultures stories are fluid and can be adapted to suit setting and circumstance, in effect they are timeless and cyclical, like the sunrise and the seasons, but in literate cultures stories became linear, they had a beginning and an end, which could clearly be discerned on the parchment. Misinterpretations and reinterpretations of these 'scriptures' led to all kinds of conflict and misunderstanding.

The Jewish Bible was originally written with consonants only, acknowledging that the vowels (the voiced sounds) were too sacred to be committed to paper. They could only be interpreted and understood by an initiated priesthood. When Jewish and Christian scriptures finally were written down in full then misinterpretations of these texts caused untold schisms and wars, as each sought to impose their own interpretations upon the written words. Meanwhile the original esoteric meanings of these texts became lost as people interpreted the texts ever more literally.

Mystics such as the Druids understood that as soon as you wrote something down you diminished it and opened it up to misinterpretation. That is why, despite having their own writing system known as ogham, they never wrote down any of their mystical knowledge. So when the Druids died out their teachings died with them, still uncorrupted by the written word.

As well as being responsible for religious divisions, wars and other forms of misunderstanding, we now also have lawyers making a fortune from literally interpreting dense and impenetrable law codes that the everyday person has no chance of understanding. It is literal interpretation of writing that is responsible for all this separation. It is knowledge devoid of all human emotion. It is the *word* of the law, not the *spirit* of the law.

'How can we limit misunderstandings generated by the written word?'

As you seek to pursue your own spiritual path to personal development you can learn from written teachings and take on board only what feels appropriate and relevant for you. There is no need to follow the written word to the letter as

if it were scripture, nor is there any need to convince others of our beliefs and get them to follow one particular spiritual path. Scripture was created to codify and fix a set of beliefs and then to impose it onto others; this only creates separation from those who do not share our beliefs and should be avoided.

If we do not want to be misunderstood then we should communicate with each other in the way that nature intended, through speech and emotion and body language. So much strife is caused in our world by things being taken literally. Just think about that word 'literally.' It means something that has been written down, as if the literal truth was somehow more important than our own intuition, feelings and emotions.

Free yourself from what has been written! It is not the truth; the truth lies within you and within all of nature. You are a living, breathing part of this world and you were not created to have laws and scripture imposed upon you by those who think they know better, you were created to be free and to realise your fullest potential as a sovereign being.

'How is it that trusting our intuition and doing what feels 'right' evolved into being told what is right by strange symbols printed on paper?'

The answer is *fear*. We no longer trust that the Earth will provide for us, now we want it confirmed, in writing! We pass responsibility for our behaviour onto law makers and religious authorities, no longer trusting in our own ability just to do what feels right. We allow our governments to dominate us and impose laws on us. We do not question that what has been 'written' may be wrong, that laws and restrictions were codified in order to control society.

'Some of the most ancient writing in the world tells of shining beings that came down to Earth to teach us laws, writing and civilisation. Is there any truth to these stories?'

One of the world's most ancient literate cultures, that of Sumer in Mesopotamia, describes such encounters in great detail on many hundreds of clay tablets that were unearthed in the last century by archaeologists. They described how a race of beings known as the Anunnaki, came down from 'on high' and taught mankind civilisation and established kingship on Earth. These beings, the 'Shining Ones' as they are sometimes called, are claimed by these writings to have bred with our ancient ancestors and taught them civilisation.

I am not here to tell you what to believe, only to reveal to you the spirit world of our ancestors so that you can absorb from it what feels appropriate to you. Whether there is any truth to these stories I cannot say, but many of the world's ancient cultures, both primitive and civilised, have stories of beings who came down from the stars to teach them.

But we as human beings are all born of this Earth, our Mother, who has nurtured us since the dawn of time, since long before these star beings were ever said to have set foot on this planet. We are a part of this world and we should take full responsibility for it and for our actions here. Such stories are little more than a distraction from this great responsibility.

12

Monotheism

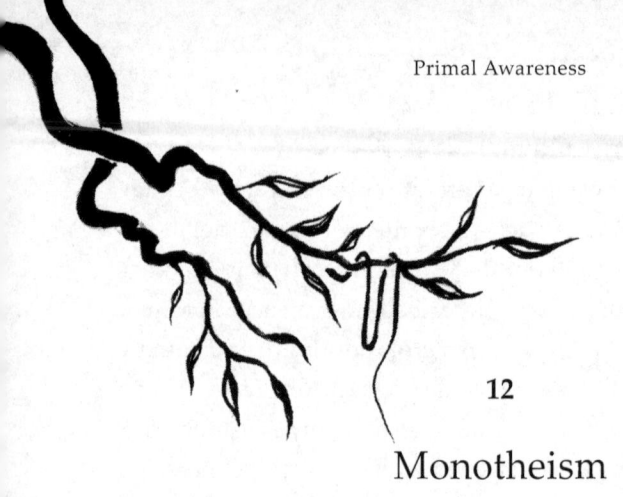

Some religions followed many gods, while others worshipped but a few. Some gods came to be seen as more powerful, and were given special reverence. These became the tribal or national gods of whole cultures. These male-dominated sky gods came to be seen as increasingly separated from nature, living detached from Earth in some heavenly realm. They projected their power down onto the land and meddled in Earthly affairs.

One of the gods who rose to prominence was the Sun God. Under the Egyptian pharaoh Akhenaton, the Sun God was known as the Aten, only this time it was different, Akhenaton declared that Aten was the only God that could be worshipped, that Aten was the one true God. It was the first recorded case of religious suppression at a state level. Until that point, religion had been inclusive, so that in addition to the state gods, people were allowed to worship any gods and spirits that they wished. But henceforth religion would become exclusive, allowing for the worship of only one God. The cult of Aten soon passed away but the idea of one all-powerful God would in time re-emerge through Judaism and through its offshoots such as Christianity and Islam.

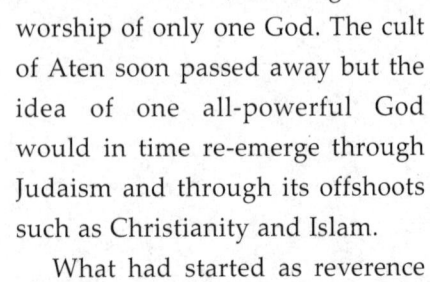

What had started as reverence for one god in preference to others, evolved into reverence for one God alone, which allowed for the worship of no others. These others were now no longer seen as gods but as lesser beings – demons and devils – their powers diminished or denied.

Monotheism, the belief in one all-powerful God, shifted people's perception of spirituality. Instead of consulting many spirits people now connected to a more abstract sense of a divine power, the belief in a single all-powerful deity. Previously people had felt that all of nature was infused with spirit force, but now this all-pervading force was detached from nature and relocated to some heavenly realm that was separated from this Earth. But this new spirituality was of a very different kind to the primal nature-based spirituality of their ancestors. Instead of consulting with spirits and gods of nature for their aid, the people were now told there was only one God and He had to be worshipped.

This was far removed from the primal concept of a radiant spirit in the sky who shone down upon the Earth freely giving forth its light and energy. The Earth Mother was denied and all of Nature was now subservient to the one true God. The Earth was His creation, created for mankind for us to use and exploit as we wished. Nature was once seen as the divine made manifest, in essence nature was God, but now it was instead seen as the creation of a divine will, the creation of an all-powerful transcendental God. Nature had become subservient to God, and to mankind, in this newly male-dominated world. The power of Mother Earth and of all things feminine was denied.

This idea of a transcendental male God was already well developed in Judaism when Christianity came along based upon the teachings of Jesus Christ. This new religion incorporated elements both from Judaism and from Graeco-Roman mystery traditions. Jesus Christ came to be seen as the son of the one true God, and also as God in human form, an earthly manifestation of the heavenly divine presence.

Islam was also a development from the ideas of Jewish and

Christian monotheism, but unlike Christianity it denied the divinity of Christ, and saw God or *Allah* as essentially formless and forbade any representations of Divinity.

Belief in only one God by necessity meant the discarding of belief in all the other gods. These gods, who originally were the powerful spirits of nature, were either denounced as being entirely false, or they were demonised and turned into devils. Other gods and goddesses were Christianised and their stories were replaced with the deeds of saints and angels. Thus we find a spring dedicated to the goddess Annu transformed into St Anne's well, or a temple to Apollo replaced by a church to Archangel Michael. People had believed in these gods and spirits for millennia, since the very beginnings of their conscious development, so they could not easily be told to simply discard those beliefs when they knew with all certainty that these spirits were real. It was much more effective to convince the people that these gods and spirits were actually demons, sent by the Devil to fool them, or that the healing properties of a holy well were conferred upon it by a Christian saint according to God's will. This made the transition from paganism to monotheism much easier for the common people to accept.

The religions of monotheism (which means literally 'One God') first caught hold in the cities where people were already separated from nature. Those living in the countryside, far from the corrupting influence of the cities, had a harder time abandoning their ancient beliefs and so the word 'pagan', which literally means 'dweller in the countryside', came to be associated with a belief in the old gods.

People, especially in the cities, had come a long way now from their original nature-based spirituality where they roved the land in communal groups hunting and gathering and calling upon the spirits of nature for aid. The rich had become disconnected, depraved and corrupt, while the poor had become enslaved and tied to small plots of land. A new kind of spirituality was now imposed upon the people which better suited the needs of a more organised and complex hierarchy.

As the Church became increasingly absorbed with its own power and wealth, and veered further away from its original spirituality, it felt the need to persecute anyone who did not agree with its dogma, and who threatened its stranglehold on power. Pagans and heretics were relentlessly persecuted and cruelly tortured and killed for their beliefs[2]. This savage persecution of their fellow humanity culminated in the crusades, the witch trials and the Inquisition of later centuries, as the Church sought to impose the will of the one true God upon all. The God of forgiveness of the New Testament seemed to be losing out to the vengeful Jehovah of the Old Testament!

There were also those who were devotees of the one God but who did not subscribe to the idea of the hierarchical Church acting as intermediary. These were the Gnostics of early Christianity, monastic orders such as the Culdees of Celtic Christianity, and the Cathars who were a later medieval resurgence. They believed that individuals could each make their own personal connection to God and could achieve their own salvation as they attained a form of enlightenment that could be called 'Christ consciousness'[3]. These people were branded as 'heretics' by the Roman Catholic Church. The Cathars and Gnostics were ruthlessly persecuted and exterminated, while the Celtic Church was co-opted and made subservient to the Church of Rome.

It took the Church many centuries to stamp out pagan beliefs and even then they were not entirely successful. Oppressive laws were passed again and again forbidding people from offering at pagan altars, asking for the help of spirits, or praying to 'false idols'. Punishments became increasingly severe as the Church sought to tighten its grip; people were threatened with eternal damnation if they consorted with these devils and did not follow the teachings of Jesus Christ. But still some would secretly visit nature shrines for help and healing, making offerings to fertility gods, or engaging in pagan rituals.

'So what is truly meant by this concept of 'God'?'

In primal spirituality, everything is seen as having a spirit. There is the spirit of the tree, the spirit of the forest, and the spirit of the land, each one contained within the greater whole. So that in the same way that a cell is part of your hand, and your hand is part of you, so a tree is part of a forest and the forest is part of the land. Each has its own identity and yet is part of something greater. This something greater, which ultimately we call the universe, itself has a spirit, which in some cultures is known as the Great Spirit. This Great Spirit is the all-encompassing spirit force that pervades all things, and that all other spirits are but a small projection of. It is the source from which all things flow, the formless that gives form to the whole world, the sky, the stars and the universe.

You could say then that God is the Great Spirit, who is the spirit of all things, the spirit of the cosmos. But this concept of a great spirit, where it existed, became confused with the Christian concept of a male God who lived up in the heavens. This monotheistic concept of God was detached from Earthly existence and projected His power down onto the Earth, rather than being an all-pervasive spirit force which is immanent within it and within all things, including us.

In the East this Great Spirit is known as the Tao, the mysterious and unknowable force behind all creation, but in the West we have this much more confused representation of this divine essence. We call it God and imagine it as a being which is separate from ourselves, rather than it being the divine essence which we are all a part of. But it is hard to imagine a worse word than 'God' to represent this divine essence, descended as it is from the image of an old pagan sky god representing male dominance.

What about the female? All is a manifestation of the divine

source, and without the interplay of male and female forces there can be no existence in this world. For our world is one of duality, of a constant shifting of energies back and forth, creation and destruction, one is not possible without the other. Yet the Great Spirit is timeless, unchanging, an immutable force that just Is, the source of all things, the very framework upon which this universe of ours hangs. Modern scientists would simply refer to this as the 'laws of physics' but God is not just the laws of physics, for the laws of physics themselves are a manifestation of divine perfection and harmony (such as can be witnessed in sacred geometry).

This Divine Source is not 'out there' somewhere, in some transcendent realm in the heavens; it lies within us, and within all things. The potentiality that created the universe exists as a blueprint deep within every single one of us and within every cell of our body and every speck of dust in the universe. It is contained within every single atom and within all things. You are God, we are all God, there is no separation, we are all part of the One.

'So how can we restore the balance between the masculine and the feminine and bring our spiritual lives back into harmony once more?'

In prehistoric times we used to revere the Earth Mother above all else, as she was the one that sustained us and gave us life. We could feel her energies coursing through the Earth and these were often represented by serpents; so in the matriarchal religion of Minoan Crete for instance we see many images of the Earth Mother holding or wrapped in serpents.

When the patriarchal religions came along, based upon sky gods, they suppressed reverence for the Earth Mother, and so we get stories of the Greek and Mesopotamian gods and heroes fighting and slaying serpent-like creatures which are associated with the Earth.

In Medieval times these serpents were known as worms, and

later on as dragons, and so we get images from those times of the Archangel Michael suppressing a dragon, or St George defeating a dragon, or the many stories around the countryside of brave Christian knights slaying dragons.

The dragons represented the old Earth-based pagan religions and the energies of the Earth Mother, a power completely denied by the Christian Church. These were often represented instead by an evil serpent or a devil, as they sought to demonise and suppress all reverence for the divine feminine in nature. So in these churches we often now see images of Archangel Michael suppressing the dragon, literally standing on it and subduing it with a spear. Once you understand the symbolism it's obvious what these images really mean. The old religion of the Earth Mother was suppressed and subdued, Christianity was triumphant.

Places where these Earth energies are strong have been revered as sacred sites since ancient times; temples would have grown up there and people would have come to these special places to connect with the potent spiritual energies of the Earth. When the Christian Church came along these temples would have been destroyed and replaced instead by churches. The early Church knew that these places were powerful and revered and they wanted to tap into the sacredness of these places. They understood that these special locations could have an influence on human consciousness, and so many of these locations were co-opted by the Church as they sought to influence the hearts and minds of the people.

The time has come now for these sacred places to be reclaimed. No longer does the Church have a monopoly on our spiritual lives, so no longer should it have a monopoly over these sacred places. They should be open for all to come and make their own direct connection to spirit, whether they believed in the divinity of Jesus Christ or not. These holy places could be utilised to help to bring in a new consciousness, one that is based around love and unity, instead of power, control and separation. This does not mean going back to simply worshipping the Earth and the Goddess, because that would be equally unbalanced, it means being accepting of all the energies of the Earth and the cosmos, both the masculine and the feminine and allowing a new spirituality to emerge that is open and available to everyone, so that we can each make our own personal connection to a sense of divinity, in whatever way we choose. Respecting the divine feminine and treating the Earth as sacred again will be a giant step towards ensuring that we no longer simply abuse and exploit it.

Currently we build structures with no thought for their harmonious placement within the landscape; they are simply imposed upon the land, creating all kinds of discord in the natural environment. In China they practiced Feng Shui in order to maintain harmony and in the ancient Middle East and Europe structures were built in accordance with sacred geometry, but even these compromises have now been forsaken as we churn up the face of the Earth with no thought for the damage we are inflicting upon the natural environment.

The human mind created the society that we live in now, and if we want to we can use our minds to shape an entirely new vision for the future. We can reactivate our appreciation for the sacredness of the landscape and the sacredness of all nature, bringing it back into conscious awareness and perceiving all of nature as being infused with a divine spirit force. Once we start perceiving the Earth and all of nature as sacred again we will then automatically feel compelled to stop abusing and exploiting it, and thereby avert our current course to destruction.

'Is there no future for religion then, if religion is the cause of so much separation?'

Whether we believe in one spirit, or many spirits, or no spirits at all, the important thing is that our beliefs be

inclusive and not exclusive, that our beliefs are tolerant of all other beliefs and welcome them and embrace them. Exclusivity, by its very nature, creates separation.

If today's organised religions can become more accepting of other faiths, then there may yet be a chance of harmonious co-existence in this world. So long as your beliefs do not impose upon the freedom and wellbeing of others then you should be free to pursue any spiritual path that you choose.

You are part of something greater, a greater consciousness that is the spirit of all things, a timeless and immortal spirit that manifests itself in the infinite varieties of life and form that we can experience here on this Earth. There are many different paths that can be taken to experience connection to this divine source, and they should all be respected. Focus upon connection, upon becoming one again with your source and the source of us all, for we are all one.

13

Reformation

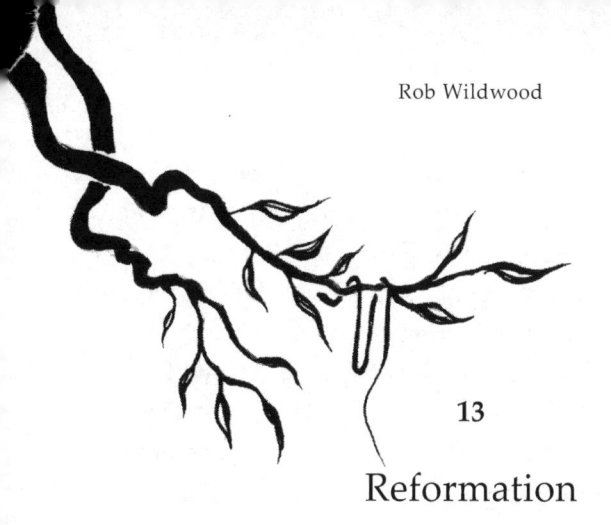

By Medieval times it was assumed that the entire population had become Christianised, but despite the persecutions of the Church many pagan practices did still survive, and continued on with a thin veneer of Christianity. Holy wells were still visited for cures, as were other sites such as sacred stones, caves, or holy mountains. Only now people would pray to the local saint instead of the local spirits. Saints had replaced gods and spirits in the popular imagination, although in many ways they still served the same function. People would visit the shrines of saints to receive cures and blessings from their holy relics. In these places they felt a direct connection with the saint, just as they would have felt a direct connection with the gods and spirits of old.

The Protestant Reformation put an end to all this, as all such practices were denounced as pagan superstition. Holy wells and sacred relics were destroyed and people were forbidden from worshipping at such 'pagan' shrines. The only appropriate place to worship now and to ask for help and guidance was in the church, praying to God. Anyone showing the slightest inclination towards pagan beliefs or practices, magic or witchcraft, could be tortured until they confessed to dealings with the Devil. Condemned witches and heretics were then publicly executed, either by burning or hanging. The separation of the people from the spirits of nature was now complete as all such beliefs became branded as witchcraft and devil worship, and so through fear of recriminations and fear of the devil people avoided such practices, and so they faded from memory over the generations.

The Protestant Reformation did have a positive aspect though, it saw a weakening of the strict Church hierarchy as people now felt free to set up their own churches and pray to God in a way that felt more appropriate for them. But they were now no longer connected to the spirit world, only to the abstract idea of a transcendental God, utterly disconnected from Nature.

'Where now can we go to reconnect with the spirits of nature?'

Fortunately, many holy wells and sacred sites do still exist, especially in the Celtic lands and in indigenous lands throughout the world. These include stone circles, burial mounds, holy mountains, sacred lakes, magic caves and hidden shrines. It is still possible to take a trip to one of these sites and contemplate there in silence, making a direct connection to the spirits of nature and the spirit of place (the *genius loci*). It is still possible to feel the power of these sacred sites, or to drink their healing waters, and come away feeling renewed and re-energised as we connect once more back into the spirit of the land.

If you do not have a sacred site near you then simply find a natural spring, or a woodland grove, or your own little corner of nature where you can commune with the spirits of the land. Connect with the spirits of the trees, the spirits of the waters, or the spirits of the earth, remembering that every river and water source, and every woodland and mountain each possesses its own guardian spirit which is its

very essence. The spirits of nature are all around you, just waiting for you to connect with them once more.

> *'Only when your people stop seeing the world with dead eyes, when you start venerating the rivers, the springs and the sacred groves once more, will there be hope for this world. Only then will we stop fearing you and feel that the time is right to open up and share our wisdom with you.'*

Making a direct connection to the sacredness of the land is not only vitally important, it is our sacred duty.

14

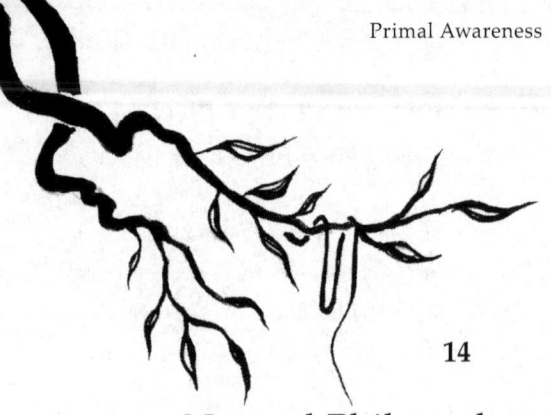

Natural Philosophy and Science

Despite the policies of oppressive religious hierarchies, the ancient mysticism never completely died out, the knowledge was kept alive by alchemists and mystics, Rosicrucians and Hermetic orders. By Tudor times this ancient knowledge was being revived and looked at with a new sense of enquiry and a new philosophical outlook that eventually would lead to a scientific revolution.

This period was known as the Enlightenment as certain educated people sought to gain a deeper understanding of how the world really worked. This movement had started long before, with the Greek philosophers and the Egyptian hermetic mystery schools, and was carried through the Middle Ages by Islamic scribes and Christian monks (including the Knights Templar), before developing into a movement known as Rosicrucianism. But this budding knowledge was constantly being suppressed by the Church as it tried to maintain its stranglehold on knowledge and power. Rosicrucians and alchemists risked their lives by challenging the Church-held view of reality.

But these natural philosophers were not scientists as we would think of them today, in fact the word 'scientist' was not coined until much later, in the nineteenth century. They were more like magicians who experimented with arcane arts in order to try to reveal the secrets of creation. Most of them still believed in God, or a spirit world, and they combined logical and intuitive thinking in their scientific work, as well as a great deal of mysticism.

The combined efforts of these natural philosophers and alchemists would eventually lead to the creation of scientific institutions such as the Royal Society, which still exists today, but the founders of these institutions, such as Isaac Newton, Francis Bacon and Robert Boyle, were first and foremost alchemists; the majority of the work they produced we would today consider alchemy, not 'serious' science. However, it is their more scientific treatises that left their mark on society and made them famous, so that they are still remembered to this day. Many of these treatises are now seen as the bedrock of our modern science. Further separation would have to occur though before modern scientific thinking was born, which would eventually come to see the entire world as mechanical and materialistic. Anything spiritual or mystical would come to be denied as the scientific elites eventually sought to impose their own strict dogma upon our view of reality.

This belief in rationalism and the scientific method would lead to incredible new technologies and propel mankind forward into an age of material abundance that was previously unheard of. Mankind would become increasingly focused on work, money and possessions, as spirituality took a back seat in a new age of materialism and rational thinking.

Nowadays we don't question that 'reason' is a good thing, but what does it really mean? It means the rejection of our intuition and our feelings in favour of only that which can be scientifically tested and proven. This is known as *positivism*. That which cannot be proved is no longer treated as mysterious and unknowable; it is simply treated as untrue!

In our primal state we accepted our natural environment; everything was seen as a mysterious manifestation of spirit, a magical experience that we entered into from birth. But now we wanted to know how everything worked. We wanted to take things apart and see what made them tick. This is known as *reductionism*, the belief that an object is nothing more than the sum of its parts. Despite the obvious absurdity of such an assertion in the case of complex and integrated systems, such as the human body or even planet Earth (Gaia), reductionism still plays a large part in scientific thinking today. This sense of enquiry has helped us to understand the world better and to create new technologies, and these new technologies have in turn added to the quality of our material wellbeing, but always this seems to be at the expense of our spiritual wellbeing as technology only further separates us from nature.

The scientific method has its inherent flaws; it can test only the material world and not the spiritual world which gives it form and meaning. Imagine if we were testing the images on a television screen as if they were our reality. We must acknowledge that it is the wiring and electronics behind the screen which give the images on the screen their form. But there is also a deeper mystery of where these images originate and what the true meaning of them is. Can we determine the plot of a movie just by analysing the pixels on a screen? There are some things that can never be reduced and accounted for using the scientific method. Science can help to produce the technology but only we can produce the human interactions, feelings and emotions that are involved.

The natural philosophers were not restricted by these limitations that scientists impose upon themselves today, they looked at our world with an open and enquiring mind in order to reveal some of its deeper mysteries by investigating both the physical and metaphysical properties of our universe.

But a great shift was now starting to occur in our perception.

In the past we viewed our reality, our experience of this world, from the inside looking out. We were an integral and inseparable part of that which we observed. But the scientific method was teaching us to view our world entirely differently, we were now to view it from the outside looking in, analysing it as if we were somehow detached from the reality which we were observing.

Ironically this idea of looking at things and analysing them from outside, from a detached distance, was not entirely new. It stemmed from the Christian belief in a transcendent God, who was somehow detached from nature, existing elsewhere in some heavenly realm, and observing Earth and the deeds of mankind from afar. Christians also perceived the human soul as being detached from nature, a kind of uniquely detached consciousness that was given lordship over the material world, which was seen as a 'fallen' realm.

Likewise, it was believed by these early scientists that human beings were uniquely conscious and could observe nature from a detached distance, as if we were not actually a part of it. Previous to this the spirit world was always seen as a part of our world, immanent within it, a world which we were also a part of. This idea of being able to somehow detach ourselves from this world and observe it from outside would have been incomprehensible to our primal ancestors and would have seemed absurd, as it would also be by primal people today who are still connected to the Earth and see themselves as an inseparable part of it.

Modern science has maintained this Christian belief that somehow we can analyse objects and phenomena in our world from a detached distance, as if we were not actually a part of that which we are observing. This is despite the fact that quantum physics proved almost a century ago that observer and experiment cannot be completely separated; that the act of observing can affect the outcome of the experiment. Scientific dogma has today become so entrenched that people find it hard to assimilate this 'new' information. The scientific method is based upon the premise of there being an impartial observer. To believe otherwise puts the whole of the scientific method and modern scientific thought into jeopardy.

Science is incredibly important for generating new technologies and generating huge wealth and so the high priests of science, as well as their financial backers, will not contemplate any form of enquiry which seeks to upset that particular apple cart. Modern science has come to dominate the world both culturally and financially, and that alone is seen as reason enough not to tinker with scientific methods and scientific 'beliefs' which are tried and tested. We cannot ignore what we have learned from quantum physics though, and this new science will be incredibly important for our future once its full implications are more fully understood.[4]

Christianity and natural philosophy created this mental separation from nature, but it is up to us to undo that mental programming and accept that we are a part of the Earth once more. Everything we do in this world has consequences; every breath we breathe contributes to the Earth's atmosphere and is circulated and recycled by all the other beings on this planet, including animals, plants and even rocks. The Earth is a living being and we are a living part of that greater body. We cannot separate ourselves from nature; every act we perform here has repercussions for the greater web of life.

'How can I experience this sense of viewing the world from the inside looking out?'

The answer is empathy, something that is sadly lacking in modern society, but something that was ubiquitous in our primal past.

So let's take an example. Let us say you want to experience what it is like to be a wolf. What would be the best way to go about this? A biologist may cut it up into its component parts to see how it works. A zoologist may analyse its behaviour, compile data on its size, shape, movements and interactions. But can they really feel what it's like to be a wolf without having empathy with the animal that they are observing? No, there are some things that simply cannot be grasped through analysis, they can only be experienced. Almost all human beings have this ability called empathy, even if we are not aware that we are engaging with it. It is not possible for a normal human being to remain entirely detached from that which they are observing.

So to really understand the essence of what it is like to be a wolf perhaps you must actually enter the mind and body of a wolf! But how can this be achieved with the limited empathy available to modern civilized people?

As human beings we still possess this unique ability to experience things that are outside ourselves. We are actually able to project our awareness out into our environment or into other beings or objects, be they works of art, great scientific projects, or our favourite pets. It is also why we find television and movies so engrossing, because we can actually feel what it is like to be immersed *in* the movie. But television does not exercise this ability; it simply puts us into a passive trance-like state where all we are doing is receiving.

To truly exercise this ability, you could try projecting your awareness outside of yourself and into another living being such as a bird, a dog or a tree.

First put yourself into a deep meditative state and then focus your awareness *into* the being you want to experience. Imagine what it is like to actually become it; feel how it feels to have four paws, or wings, or branches reaching up to the sky. The more you focus on it, the more you observe its every tiny movement and mannerism, the more you can actually feel what it is to become it. In the past this was part of an art called 'shape shifting'. This is the ability to feel what it is like to be in another one's skin; to feel an actual connection

with another living being; to mimic it; to become it.

Practise this process often and it will help you feel a connection and empathy with another living being, and you will start to appreciate how it feels to actually *be* that being and to be immersed in its environment. Repeating this process with other beings will expand your awareness into the living environment and eventually you will have a feeling for the whole environment around you, as well as each creatures place within it, including your own. You will have developed your empathy.

This is where the true essence of something lies, not in taking it apart and analysing it, but in *feeling* your way into it. This is **esoteric** knowledge, knowledge from the inside looking out, not **exoteric** knowledge, knowledge from the outside looking in.

Science has enabled human society to be filled to overflowing with the exoteric *knowledge,* but now the time is ripe for a return to balance and harmony with nature, for empathy with all beings, and a return to esoteric *knowing.*

15

Industrialisation

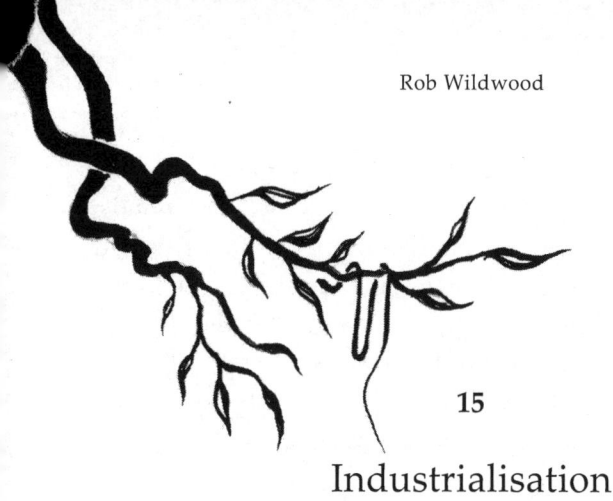

Before the industrial revolution the majority of people still lived in the countryside, working on small farms, closely connected to the land.

Belief in the spirits was still widespread amongst these country folk, only now they came to be perceived differently. No longer perceived as immanent within nature they were now seen as visitations from another world, a twilight realm of faerie that somehow overlay our physical dimension. These Otherworld beings came to be known by such names as the fairies, hobs, goblins and piskies. But these were not the cute creatures of modern fairy tales; they were oftentimes dangerous and unpredictable spirits who could be encountered on lonely trackways at night or at certain auspicious locations in the landscape.

Some of these locations could also be visited for cures, good luck, or divination, while others were best avoided altogether. Tales of these otherworldly beings were shared regularly around the hearth fire or by passing travellers, often with a hint of irony as if they were to be only half believed.

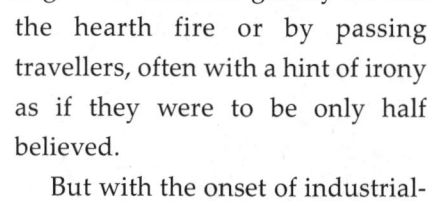

But with the onset of industrialisation more and more of these country folk started moving to the cities where their quaint folk tales

would be mocked if they spoke out about them, and so they kept quiet and the tales lay buried in the back of their minds, remaining unspoken. New generations were then born who'd never heard the tales or didn't take them seriously, and so the stories and traditions were no longer being passed down. The old folk still kept the knowledge, but they were no longer willing to talk openly for fear of ridicule, and so slowly and surely the old beliefs and practices died out and the stories faded from memory and passed beyond all recall.

The only reason we still have knowledge of many of these traditions today is because a handful of antiquarians in Victorian times visited remote rural areas and gained the trust of some of these old country folk and collected their stories. Were it not for this we would have assumed that such beliefs had died out centuries before, but it is remarkable how enduring the old beliefs are, and some have even survived to this present day as folk traditions in some remote rural areas still attest.

Today we still hunt for eggs at Easter, on the feast day of Eostre the goddess of spring and fertility; and we still decorate our homes with holly and ivy on the feast day of the unconquered Sun as he begins to rise from his winter slumber at Yule. We still feel the urge to offer coins whenever we visit a sacred spring or a wishing well, only now of course we have completely forgotten the reasons why we do these things. Once we knew why we celebrated the start of spring, or why we celebrated the winter solstice and the return of the Sun, or why we made offerings to the spirit of the well, but all such meanings have now been lost to us.

Industrialised farming put a final end to mankind's close connection to the land, no longer were people working small plots of land for their own subsistence, reliant upon the forces of nature, but instead farming was passed over to land managers who farmed huge fields of monocultures with emotionless machines to feed the ever-growing populations in the cities.

Farming itself became part of the industrialised machine focused only on making a profit for the producers.

Smallholders were an ever-dwindling minority; their descendants no longer listening to old stories but instead listening to radio or watching television and having their minds moulded by what they saw and heard on there. They moved away from the countryside to the cities, intent only on finding work and joining in the urban social life. The old beliefs and traditions were forgotten and were replaced by a more modern 'popular' culture.

The focus was now on work, money, and material possessions. Spirituality became associated only with dusty old churches and droning priests. The people had forgotten completely their personal connection to the spirit of the land.

Via the media we are now told repeatedly that the only truth is the scientific truth, and so with our connection to spirit completely severed what else are people to believe? For many, any kind of religion or spirituality has become a thing of the past, something to be mocked and ridiculed, or something fit only for New Age hippies.

The media carries the beliefs of the age and we are being brainwashed by being constantly subjected to the opinions of so-called 'experts' who are themselves spiritually dead. But we can still find that spiritual connection, simply by switching off the television and our electronic devices, and going out into nature for a while. The sacred landscape is still there and the spirits of nature are out there waiting for us to connect with them once more.

'How can I make a connection with the spirits of nature?'
Find a location in nature that feels special to you and make a small shrine there where you can make offerings to the nature spirits in order to connect with them once more. Use only natural materials that will easily biodegrade. Offer food items (dried fruit, nuts, berries, honey or cream etc.),

or liquid placed in a small bowl (milk, mead or beer), or make a natural offering from handwoven grass, vines or reeds.

Place these items with a sense of reverence and ask the spirits of nature to connect with you. Repeat this process and with patience and an open, calm and receptive state of mind you will soon start to experience the magic of nature and the magic of the spirits that dwell there. By opening the door in this way you allow magic to enter into your life and you allow the possibility of the spirits of nature connecting with you. Too long have spiritually dead 'civilized' people walked this Earth now with closed minds.

'So what exactly are these beings from the 'Otherworld' that some people claim to have seen?'

Throughout the whole of history and in every culture of mankind we encounter stories about the spirits and beings from the Otherworld. These include the Scandinavian elves, the Irish sidhe, the French fae, the Arabic djinn, and the Hawaiian menehunes.

These were beings seen only in the twilight, in remote rural areas. They had a form at once enchanting and disturbing. Beings that our rational minds tell us have no right to exist.

There are thousands of stories of encounters with these Otherworld beings. Are they the spirits made manifest? Are they beings from another world who pass only briefly through ours? Are they the product of second sight when suddenly it is possible to see through the veil into another dimension? Or are they perhaps wildly embellished tales made up by the mind to account for phenomena that it cannot easily comprehend in a rational way?

Perhaps they are all of these things or perhaps there are other explanations, but there are many dimensions to the world of spirit, and the vibrational resonances that are made manifest to us in our reality are but one of the infinite and unknowable worlds that constitute the great mystery of our universe. Quantum physics is now showing us that there are many other dimensions of reality, so unexplained phenomena should not simply be discounted as delusions just because they do not fit into our currently restricted worldview.

It has been shown that our brain acts as a filter, filtering out the constant stream of information that we are being bombarded with, and delivering to us only the information that we need in order to function in this world. So much is hidden from us because our minds are closed. These Otherworld beings, if they exist, are clearly not of our simple 3D physical reality, and so cannot be tested with our current technology and physical instruments, and so remain unknowable and mysterious.

The multi-dimensional realms of spirit surround us constantly but we are completely unaware of them, tuned in, as we are, only to our own limited version of reality. Show a child an empty box and ask her what is in it and she will say 'nothing'. But we as adults have learned that the box is full of air; helium, oxygen, carbon dioxide etc. Then we learn that there are other invisible and mysterious things passing through the air; radio waves, light waves, gamma rays, the whole electro-magnetic spectrum. There is radiation, electricity, sound waves; layer upon layer of invisible forces, and there is so much more that is yet to be discovered. We live in a universe of infinite complexity containing many invisible forces that have yet to be revealed to us. Quantum physics is only just beginning to reveal these other dimensions and who knows what may one day be discovered there?

Remain open minded, accept that we will never know everything there is to know, the universe is full of mystery, honour that mystery and accept its presence within your life.

16

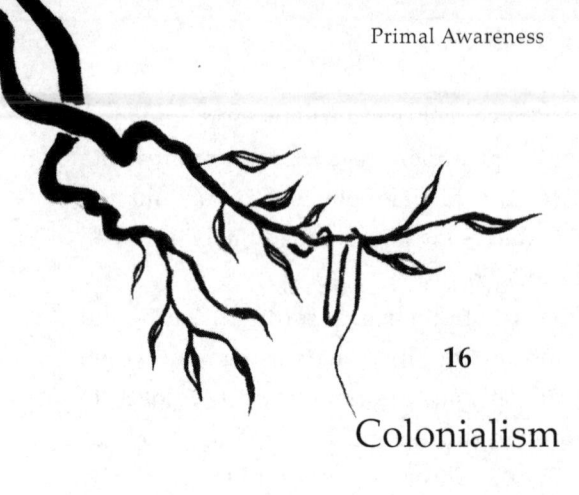

Colonialism

During the colonial era huge swathes of the world were put under the control and dominance of Western civilisation, in fact it was only a handful of countries that escaped this trauma.

Tribal people were displaced, their connection to the land was severed, and their shamans and elder wisdom-keepers were persecuted and killed. Peaceful egalitarian societies had hierarchy and patriarchy imposed upon them. Their culture was destroyed as their minds became indoctrinated with monotheistic religion and consumerism.

People living simple traditional lives connected to nature were seen as primitive by the Westerners, and in their arrogance they saw them as being in need of help or salvation. They imposed their power over these 'primitive' peoples and made them envious of their material culture and dominance. They placed no value on indigenous cultures and beliefs, on their connection to nature, their sense of community and purpose, their mutual support networks, their feeling of spiritual connection, and their reverence for the great mysteries of nature. These indigenous people were not even aware that they were poor or lacked anything until they were dragged into our moneyed society on the very lowest rungs of poverty and then made to pay for our mass-produced goods. We made these people feel foolish and primitive. They saw only our success and material wealth, and soon they envied it.

Surely the gods must have favoured the Western people to bring them such abundance? But they could see that we were hollow inside spiritually, empty and disconnected. Of all the cultures in the world it was Western civilisation that had become the most separated from nature, and so it was a cruel irony that it was the values of this Western civilisation, with its material-based culture, that were foisted upon the rest of the world.

The young ones envied our way of life and wanted their share of it, but the elders and the wisdom-keepers knew better, they could perceive what we lacked, and could see that spiritually we saw with 'dead eyes'; we had lost our connection to the land, and we had no idea what our purpose was here on Earth. The elders feared us, our spiritual deadness seemed completely alien and incomprehensible to them, and so they kept their knowledge secret, passed it down only to those who they could trust, and awaited the day when their knowledge would be valued once more.

We live in a world today where the examples we can draw upon from traditional societies are diminishing daily as rampant capitalism fuelled by globalisation spreads to every last lonely corner of the globe. Natural resources are exploited and native people are drawn inexorably into our moneyed economy.

Where now can we find the inspiration for living a better life, one that is more connected to the land and to the spirits of nature? Without these examples from indigenous peoples who are still connected to the land and its spirits we are lost.

'So what can be done to protect the remaining tribal people and preserve their culture?'

We can try to protect the few remaining tribes through direct activism or by helping charities that have been set up

for this purpose, such as Survival International or The Mother Earth Restoration Trust. That way we may play our own small part in helping to preserve their ancient culture.

You may also benefit from studying indigenous cultures, to enable you to come to a better understanding of their way of viewing the world. It is very different to ours and will give you an entirely new perspective on life (see Recommended Further Reading section at the end of this book).

Modern travel and communications have shrunk the world to such an extent that it is easy today to believe that all the world's mysteries have been revealed, that there are no more hidden places of magic where it is possible to live in a different way to ours, one which is so at odds with our way of viewing the world. But the old magic is still there, and if you go looking for it and explore deep enough, and allow yourself to be carried deep into the heart of nature, you will find it there.

17

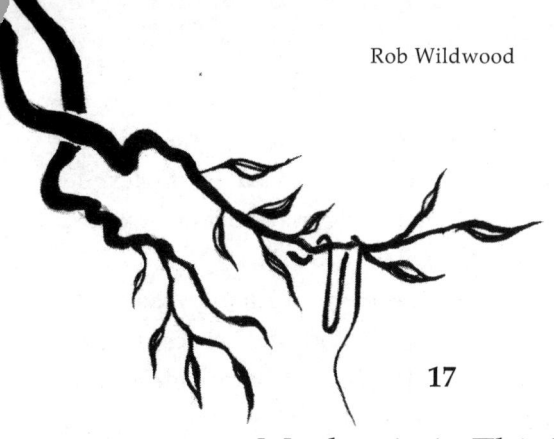

Mechanistic Thinking

Christianity came to view our universe as a giant clockwork engine, set into motion by God. Only human beings had souls they believed, all else were just machines, placed there by God for us to use.

This mechanistic thinking carried through into modern science where everything came to be viewed from this detached perspective. Physics, chemistry and even human origins came to be explained in these purely mechanistic terms. Only we had souls, only we had consciousness, therefore only we could take a perspective that was detached from the machine and observe it objectively.

Theories about evolution developed from some simple observations of nature. Firstly, that species have adapted to fill niches in their environment, and secondly that the weak will die while the strong and better adapted will survive to pass on their genes. Such an observation is simple and self-evident, and is not in dispute.

But this neat theory of evolution espoused by Darwin and others (who were all Christians) was taken to extremes by the so called 'Neo-Darwinists'. They claimed that the whole universe was simply mechanistic, and that everything, including evolution, happened purely by chance. They denied that there was any mystery or mysterious forces at work, despite the fact that there are so many deep mysteries to the universe that science has never come close to explaining.

For example: how does life coalesce from simple chemicals? What is the force that gives and sustains life and consciousness? How is it that purely random mutations are able to achieve such perfection? What created the laws of physics in such a precise way that our world is made possible?

Such mysteries are unanswerable in the mechanistic mode of scientific thought. To claim that the Earth is just one big machine that was thrown together by random chance should now be a completely outdated way of thinking. It numbs our mind from the ability to conceive of something greater, some force greater than we can comprehend that is a guiding principle behind our reality, the Great Mystery as it was once called. To be cut off from this sense of mystery is to be in a spiritual wasteland.

Spirituality is based upon mystery, upon mysticism, upon accepting that there are forces beyond our control which we cannot comprehend with our rational minds. To be removed from a sense of mystery is the worst kind of separation there is, it creates disillusionment and depression, and a sense of isolation and pointlessness. Are we really just randomly created animals struggling for survival so we can pass on our genes? Of course not! Science does not need to be tied to such mechanistic ways of thinking and outdated ideas, it can open up to new possibilities, so that science can once again serve the sense of curious and joyous enquiry that it was always meant to.

It's important to acknowledge that there is something greater than us, something greater than our knowing, greater than our five senses can ever relay to us. The Great Mystery is all around us every day. It causes the Earth to turn and the Sun to shine, it causes plants to grow and winds to blow, it breathes life into our bodies, it animates us and sustains us. It is the mystery that created all life, and that enables the great Mother Earth to constantly generate new life that adapts to and fills every part of her vast nurturing body.

No creature lives in isolation, no creature struggles alone for its existence and survival, for all are part of the whole, part of something greater, something that is the guiding force that generates life and consciousness. Just as the cells and organs in your body cannot live without you, so every plant and animal on Earth cannot live without its mother the Earth, for we are all a part of her, there really is no separation; it only exists in our minds and the strange mental constructs we have bought into, such as our mechanistic thinking.

Do the cells or organs in your body fight one another? No, they get along as part of one harmonious whole[5]. Likewise, the cells in the body of Gaia, Mother Earth, each have their own role to perform as part of the whole. The hunter and the hunted are one, part of the greater functioning of the body of Gaia.

Countries with a majority indigenous population have a very different worldview to us. Bolivia and Ecuador have recently amended their constitutions to grant equal rights to nature. This so-called 'Wild Law' is the result of a resurgent indigenous Andean spiritual worldview. They see Mother Earth (or Pachamama as they call her) as a living being. The law states: 'She is sacred, fertile and the source of life that feeds and cares for all living beings in her womb. She is in permanent balance, harmony and communication with the cosmos. She is comprised of all eco-systems and living beings, and their self-organisation.'

Reductionist scientific thinking would see this worldview as deluded, but it is Western civilisation that is destroying the planet, not these indigenous people. Now whole nations are standing up and saying: 'No more!' Could it instead be modern scientific thinking that has somehow become deluded by convoluted thought processes that obscure reality and prevent us from seeing the bigger picture?

The government of Bolivia has recently granted to Mother Earth and *all* her life systems, including human beings 'the right to life, to diversity of life, to water, to clean air, to equilibrium, to restoration and to freedom from contamination'. Any new development will now have to come up against these laws which have been granted to the natural environment. We may think of modern science as progressive, but it could be these indigenous thinkers who hold the real key to our future.

What is really needed at this time is an end to the old reductionist way of thinking and the embracing of a new holistic science that views systems as a unified whole and not as the mechanical sum of their parts. This has already occurred in some areas of science, such as Earth Systems Science which developed out of James Lovelock's Gaia theory. Lovelock saw the Earth as a giant living being, a complete self-regulating system that maintains its own equilibrium. This powerful theory (though nothing new to indigenous people) is now slowly being accepted and integrated into modern science.

Progress is also being made towards recognising that animals are fully conscious living beings just like us. Previously science held the Christian view that animals are just mechanistic creations that have no human-like consciousness or self-awareness. But many governments around the world are now recognising that animals are conscious beings just like us who should have rights too.

A scientific study in the UK conducted by a group of neuroscientists concluded that many animals have similarly constructed brains and nervous systems to human beings, so there is no reason to suppose that these animals are not also conscious and aware. This led in 2012 to the Cambridge Declaration of Consciousness which declared that many animals, including all mammals and birds, and even octopi, are conscious beings. This can be seen as a small step towards recognising that everything on Earth, including Earth herself, has consciousness, but unfortunately these scientists are still looking at it from the standpoint of the brain generating consciousness rather than there being a universal consciousness which we are all tapping in to. Belief in this universal life force, or spirit force, is common to all indigenous cultures, but science still has a long way to go before it will be able to reconcile itself with this core belief.

Modern scientific thought struggles with issues like consciousness, which, according to its mechanistic and reductionist way of thinking, really ought not to exist!

Nature is all-powerful and spirit force pervades the entire cosmos. We are all a part of this great mystery whether we choose to believe in it or not. Our beliefs and human thought processes are of little consequence to the greater workings of the cosmos, we merely have to decide whether we wish to live in harmony with this greater force, or to work against it and live in discord.

'I feel like I too have been indoctrinated with this mechanistic and reductionist way of thinking. How can I free myself from this mental prison, open my mind and experience a sense of the great mystery once more?'

You can gain a sense of this great mystery for yourself by simply planting a seed and watching it daily as it grows. Observe as it slowly germinates, pushes up through the soil and transforms into a seedling, and then rises and expands towards the sunlight until it develops into a fully grown plant, before blossoming into the most beautiful flowers of intricate symmetry. Take time to observe each stage of this process in detail, and come to appreciate the timeless wonder of its creation. There is some mysterious force behind this growth, behind this beauty and perfection, some mystery that we cannot simply explain with our rational minds and reductionist thinking. There is some guiding principle that is manifesting itself in this form at this moment for us to experience with our senses.

As the plant blossoms it is communicating with us and with the outside world, it broadcasts its beauty and insects respond to the call by coming to collect nectar while fertilising the plant. A symbiotic relationship that all benefit from.

In a similar way, a fruit tree proudly displays its fruit to the world, communicating with the more mobile species, saying, 'Come and eat me!' so that its seeds might be spread out over the wider world. Is the tree just a dumb machine, robotically dropping fruit? No! It is vitally important for the future of our planet that we stop perceiving it in this way. It is a conscious living being, and it is communicating, with us!

In time the seeds will produce more plants to complete the endless cycle of ever-sustaining life. To fully observe and engage with this process is to be filled with a sense of wonder. As you observe the emergence of the delicate structure of the leaves, or the beauty of the blossoms, you are directly experiencing a manifestation of the Great Mystery itself.

This mystery is something beyond our comprehension, it is beyond our simple science to mimic or recreate. A being of infinite complexity transforms before our eyes, a self-sustaining organism that has existed on our planet for millions of years; millions of years before mankind ever walked this Earth to tend to it, and yet here it still is, not a product of chance and random mutations but one of the infinitely complex manifestations of the divine mystery.

Sit in silence and contemplate this Great Mystery for a while, and all the different ways in which it can manifest itself. Then go out into nature and observe this great mystery all around you, in the growing plants and scurrying animals, in swirling winds and eddying waters, in layered rock and drifting clouds, in the stars at night and the play of sunlight through the leaves by day. *Know* that these creations are timeless, that they issue forth from a world of spirit that exists outside of our limiting concepts of space and time. Each is manifesting here, in this place, in this time, in its own unique way, for you to experience, in *this* moment!

Science and technology can tinker with nature, can bend, alter and distort it, but can never replace it.

Mankind, for all its intellect, has never come close to matching the ingenuity of this 'simple' seed. A miraculous organism that is able to grow and transform into something thousands of times its original size using only the natural elements that surround it. It is able to branch out and transform sunlight and air into matter and is able to send out roots deep into the earth to gather minerals and water to sustain itself. And more than this, it is able to produce hundreds of copies of itself, using all kinds of ingenious methods to disperse itself over the land, adapting and surviving over long ages of the Earth, for millions of years! The mechanistic technology of mankind just cannot compare to this, crude and senseless imitations of nature that last just a few years before rusting and decaying.

If we can see so much mystery contained within a single seed, then imagine the overwhelming sense of wonder that can be experienced once you venture out into the natural world with these fresh eyes. Allow yourself to fully appreciate the mystery of all the growing, blossoming and fruiting beings that surround you as they communicate and interact with one another and with you. Sense how all the vegetation that surrounds you is a miraculous manifestation of this Great Mystery, but then notice the flutter of bird wings or the buzzing and chirping of insects; mysterious beings that inhabit this magical world of vibrant vegetation. How did they all manifest here and grow together into this complex ecosystem? What mysterious force created all this beauty and perfection which surrounds you? The great mystery is

everywhere. It is all around you, always. You can experience it in the plants as they grow, in the birds and insects as they take flight, in the clouds as they sail by and in the Sun that rises each morning to herald a new day. And as you find yourself immersed in the wonder of this world remember that you too are a living part of this great mystery.

'So why did our minds develop to such a size that it enabled us to achieve far more than simply survival and reproduction? Why are we able to analyse on such a complex and abstract level?'

Neo-Darwinists will tell you that this just happened by chance, that the more intelligent ones were able to out-survive the less intelligent and reproduce themselves. But nothing in nature happens by chance, it is all part of the greater consciousness of our Mother Earth, and a manifestation of the great divine mystery. Humans are meant to be here; we are needed here, to bring this planet to a new level of consciousness and awareness, but that consciousness will not be achieved by focusing on materiality.

We have the potential now to reach untold heights of conscious awareness, but we should find a way to balance the logical and analytical (masculine) side of our minds with the more intuitive and sensitive (feminine) side to open up a pathway to our future development. A new way of living and a new way of thinking that is more in tune with the Earth and with natural cycles. We can use our minds to reimagine a future for ourselves based upon a new holistic science that produces technologies that work in harmony with the Earth and in alignment with spirit force. Our current course, guided only by materiality and mechanistic thinking, will lead only to destruction, both for ourselves and for the many species currently struggling for survival on this planet.

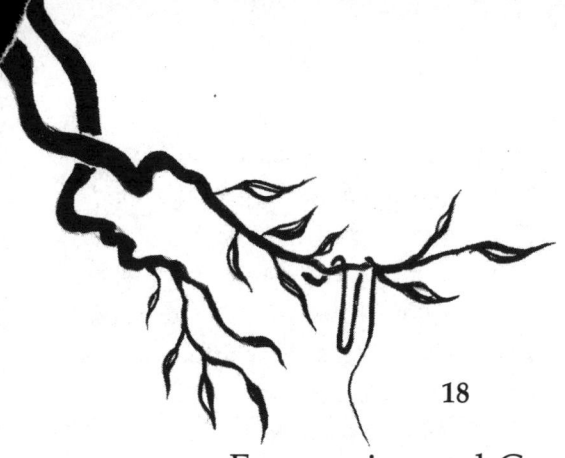

18
Economics and Capitalism

The twentieth century saw huge amounts of separation all over the planet. Nomadic hunter-gatherers such as Kalahari bushman and Australian Aborigines were forced to abandon their ancient lifestyles, live in settlements and adopt Western customs. Tribal people all over the world were subjugated by nation states and introduced to Western goods and Western lifestyles and were dragged unprepared into modern society.

Meanwhile in the more developed countries, warfare broke out on an industrial scale as scientists developed ever-more powerful weapons with which we could attack and destroy each other. On these battlefields mankind had created a hell on Earth of shell craters, rotting corpses and acrid fumes. It is hard to imagine being more separated from our spiritual source than to be on one of these battlefields where mankind had created this living hell. **This** is what our separation from nature and our mechanistic thinking had eventually led us to. The next step could well have been nuclear war and total annihilation.

It is only the human spirit and people's good nature towards each other that kept people sane throughout those horrific times, the human spirit at least had survived.

Back at home people had

become absorbed by the political struggles of a patriotic 'total war'. The nation state had risen to prominence in people's minds and so, filled with a sense of patriotism, people struggled for their nation above all else. Loyalty to the nation had taken priority over any sense of connection to fellow human beings around the planet, or to nature as a whole, as we churned up the face of Mother Earth with bombs and craters.

Returning from these wars people tried to build a better society based upon social justice and equal rights but they were easily distracted by the new entertainments of radio, cinema and television. Reality was no longer what they perceived with their own senses but increasingly came to be what was presented to them via the media. As people came to view the outside world via a screen instead of with their own senses this only led to a greater sense of separation.

The economic imperatives of capitalism also started to radically change people's view of the world. No longer was a forest a place where native peoples could hunt for food, build shelters and be sustained; now a forest was seen solely as a source of potential profit. Its resources could be exploited, its timber exported, its land replanted with cash crops.

The economic model of capitalism is based upon constant economic growth. The logical conclusion of this process is that eventually all the world's resources will be consumed. At which point the whole system will collapse. As resources become scarcer, supply and demand dictates that they will become more expensive. Therefore, as capitalism reaches its natural conclusion of consuming the entire planet the process will actually become accelerated as scarce resources become more valuable.

Economic growth is required by rich investors who want to make a quick profit, but economic growth is not always necessary. Japan experienced no economic growth for over two decades, but the people were doing fine and still enjoyed a good standard of living, especially when compared to less developed countries. We need to find an alternative to this constant economic growth and find a way of making the system self-sustaining. The capitalist system has served well as an economic model for the past couple of hundred years, but now it is like a rampant beast that is out of control, a new system is needed, a way of living which is more in tune with the environment and with our fellow beings here on Earth, a system that does not gobble up all of the world's resources but which is instead designed to provide for the needs of *all* the Earth's inhabitants.

Everything that mankind has ever produced is created from the body of Mother Earth, and it will soon enough return to the Earth, but this cycle should not be exploitative and short-sighted, producing endless pollution and waste. We should make it self-sustaining by altering our mind-set on a global scale, so that we honour the sacredness of the land and treat it with respect, reversing our current path which can lead only to economic collapse and environmental destruction.

'Should we all just go back to being hunter-gatherers then, living off the land, roving around and gathering wild food?'

We cannot go back to living as our ancestors did, we simply could not survive as we are no longer adapted to living a primal existence and the resources are simply no longer there. There are far too many people in the world now, and unless these figures are drastically reduced through stringent birth control it would be impossible to give up commercial farming.

But we can try to make the best of our current situation by creating a more sustainable society, one that respects the Earth and all the beings that dwell upon her, cultivating our own personal connection with the land, growing our food locally, trading with our neighbours, using renewable resources and supplying ourselves with green energy.

We were born here on this Earth for a reason, and that

reason is not to acquire some temporary sense of power and status by accumulating more possessions. It will surely be more rewarding to follow eternal principles of living in harmony with the Earth and to use the tools we have been given to raise our consciousness to new heights of awareness. We have already come a long way; we just cannot see it yet because we have become so focused on the material things in our lives and on all the conflict and negativity in the world. But a new way of thinking is emerging, it has taken root and it is growing, and one day it may transform the whole of humanity and create a better future for us all.

Be thankful for all the incredible progress that has already has been made. Less than 200 years ago it would have been unimaginable that women could have equal rights, that working class people would be allowed to vote, that indigenous people could be treated as equals, that slavery would be abolished, that pagans and heretics would not be persecuted, and freedom of religion would become total. Nowadays people in developed countries are free to travel almost anywhere they please. Even alternative healing methods such as reiki have now become commonplace as people open up to new ways of perceiving our reality.

But this is in our world, the affluent 'developed' world, a world that is still supported by the exploitation of the world's poorer countries. In many of these poor countries people are still destitute and hungry, they are still riddled with disease, and they still work in conditions that are no better than slavery. They are still incredibly oppressed by repressive regimes, women are still treated as property, religious minorities are still persecuted, and warfare and displacement are still things that have to be regularly endured.

In the West we have come a long way towards a new and enlightened way of relating to each other and of governing ourselves, and we should be thankful for that, but we should also find a new way of relating to our environment, to the planet and to the whole of creation. We need to cast off these chains of materialism that have been holding us back from realising our full potential as consciously enlightened beings.

Human society over the previous centuries, since our primal separation from nature and from spirit, has been following a curve, from connection, to separation, and now slowly back to connection again. Surely the low point was reached by the Middle Ages when women, serfs and children were completely oppressed and had almost no rights. The Church had complete dominance over our spiritual lives. People were routinely tortured and persecuted. Warfare and suffering were endemic. This was followed by a renaissance in learning and a new age of enlightenment, when new ways of thinking and relating to the world and to each other became possible. If we are to complete this cycle back to connection, we should continue to find new ways to move forward. This is not about going back to some 'golden age,' it is about rediscovering our connection to the Earth at a new and higher level of conscious evolution.

You can use the tools you have learned to meditate upon these issues and see if you can gain an insight into new ways for the human race to move forward in the future. No one person has all the answers, but each person has something unique to contribute.

19

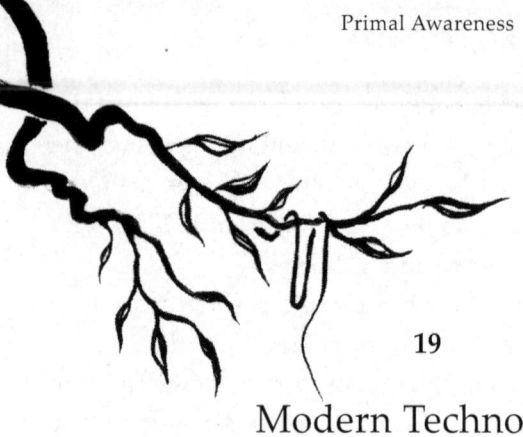

Modern Technology

Our most recent separation has come through the use of the internet and mobile devices to connect with one another. We think that we are connecting using these devices but in reality we are more separated than we have ever been. Instead of having real human interaction with speech and body language, and with feelings, touch and smell, now all we have are some words typed onto a screen, or at best a flat and sterile screen projection of the person being talked to.

We type these strange symbols known as words, we wait, and then we get more symbols back in return, which we then have to interpret using our own preconceptions about the person we are communicating with. There is very little 'real' human interaction going on here, information can be related, but feelings are often misinterpreted due to the limitations of language, and emotions are lost behind the words. Every word that is typed will have a different meaning to every person who reads that word. The true meaning is obscured, the feelings and spiritual connection we imagined whilst typing it end when we hit 'send', they rarely carry forward through the words, the message received at the other end is only an interpretation.

Of course we had this same

limitation when we sent each other letters, but letters were used far less often, and it was understood that a letter was something special and would be written in a special way. But with the internet and mobile communications it is something we are doing all the time as if it was normal speech, but it is not. It is lacking the majority of the content of normal speech, it is lacking the feeling and expressed emotions, it lacks the inflections and emphasis on the words, and it lacks normal human interactions and shared experience. When using our devices, we are not connected to our friends we are separated from them; an artificial electronic barrier exists between ourselves and those we are trying to connect with.

Text is useful for exchanging information, but not for feelings of real connection, for any connection imagined will be little more than self-projected thought forms expressed in type, and any connection that is felt or experienced is rooted in our own mental interpretations.

We spend our time now driving around now in metal boxes instead of walking on the green earth, we work in confined spaces surrounded by mass-produced electronic devices instead of crafting from nature in the open air, we confine ourselves in air conditioned cubes and gaze transfixed at moving images on an artificial screen instead of lighting a fire under the stars and listening to the sounds of the night.

Is it any wonder we feel so disconnected and stressed, so tired and sick? We live our lives now vicariously through a screen, watching what others do and where others have been, instead of experiencing these things for ourselves and experiencing our own connection to the Earth.

'But doesn't technology and the internet connect people globally?'

The internet serves an extremely useful function of disseminating information around the world. No longer can

people with free access to the internet be kept in ignorance and oppression. New ideas are circulating fast! So when you use the internet try to use it to spread positive messages that encourage acceptance, connection and spiritual awakening. Try not to focus upon the negative things in this world, they will sort themselves out in their own good time, once a new consciousness emerges and a new spiritual awareness grows. Spread positive messages of love and inspiration, for surely then you will become an inspiration for others and love will grow.

The internet connects people electronically, but at the same time makes them more separated from physical reality than ever before. So try switching off your electronic devices for a while, lock them up securely in a cupboard or at a friend's house, then go out and find real connection in this world: real people, real life and real immersion in the natural world. Seek these things out, they are needed to heal our souls and to make us feel whole again; and to make us feel once more like we are a living part of this natural world.

Summary – The World Today

All life on Earth is ultimately descended from the rocks and the minerals which form the basis of our planet, these are our most ancient ancestors, and before them, the stars. When looked at from this perspective all beings on Earth are related, we are all brothers and sisters, children of Mother Earth; the plants, the animals and the rocks, we are one. This is why in indigenous cultures these plants, animals and rocks are referred to as 'people', it's to help us to remember, and to show respect for our fellow beings on this Earth.

In our arrogance we pitied these primitive societies because we observed only what they lacked materially and ignored the incredible storehouse of wisdom and spirituality that had sustained them over many millennia. So while we have gained materially it is we who have become impoverished spiritually.

But we were all native people once, before development separated us from who we truly are, and from our true nature. Once we too enjoyed complete immersion in nature, through the air caressing our naked bodies, the earth squeezing between our toes, the food that we gathered by hand and the Sun that daily warmed our skin. We felt a complete connection with the Earth, our mother, in every moment and were consumed by the experience of her sights, sounds, smells and sensations. We felt the joy of this connection in every moment and we felt a deep and enduring love for the Earth.

Belief in nature spirits and the gods once prevented us from destroying the Earth, but now that we set no value by such things we feel free to plunder the Earth as we please. Belief in gods and spirits may seem naive to people in modern society today, but it served the purpose of protecting the Earth for millennia and gave us a deep respect for Nature. For 95 per cent of mankind's time on Earth we believed in these spirits, they were an integral part of our lives and we evolved with them as a species. This connection to the spirits of nature sustained us through the long ages of our development, so even from an evolutionary perspective this connection must have been essential for our survival.

There was a time when we and the land were one, we were born here on the land, the land sustained us with the fruits of its abundant body, and gave us the joy of deep connection every day. When we died there we returned to that same land to continue the endless cycles of renewal which are the manifestation of all life on this Earth.

We now no longer live out in nature maintaining a direct physical contact with the Earth and with the fruits of her body, instead we clothe ourselves in synthetic fabrics, live in air-conditioned apartments in crowded cities, travel to work in metal boxes, where we engage with soulless machines and electronic devices and then purchase our pre-packaged lunches with little plastic cards.

We no longer gather around a campfire under the stars, to share stories, songs and entertainment. We instead sit isolated in our apartments, communicating via email and social media, getting our information from books and the internet, and our entertainment from more electronic devices.

We no longer live in an animate world where everything is infused with spirit and mystery, we instead live in a materialistic society where the accumulation of soulless possessions and an abstract idea of 'wealth' is seen as the pinnacle of achievement.

Over time we have become ever more disconnected. As well as losing our physical connection to nature we have also lost our spiritual connection. This started with the advent of patriarchal religion when we came to believe in a pantheon of gods who were somehow detached from Earthly existence. Belief in these gods was eventually replaced by belief in one all-powerful God who dwelt up in the heavens, looking down upon the fallen realm of the Earth, which was deemed to be a place of sin.

Further disconnection from the spirits of nature occurred when the Church convinced us not to believe in the spirits but to believe instead in the power of God and the healing powers of saints, their holy places and their relics. The Reformation then swept away all these remaining vestiges of magic and ritual relating to the Earth, leaving God as the sole source of our spirituality. Eventually atheist scientists came along and told us that God did not exist and that everything in the universe happened according to soulless mechanistic principles, so we were left with no spiritual connection at all, and we were provided instead with all kinds of luxuries and material benefits to fill the gaping hole that had been left in our spiritual lives.

In medieval times society had become calcified by its rigorous adherence to Church dogma, which was eventually swept away in a new age of scientific learning known as the Enlightenment. Nowadays it is scientific thought itself which is becoming calcified and a hindrance to our spiritual development and conscious evolution. Old ideas of scientific dogma will also inevitably be swept away in another New Age of Enlightenment that is already happening. It is time for change.

Quantum physicists today tell us that there are other dimensions and hidden worlds, and that our world is in fact shaped by our very act of observing it, but even so, our rational minds have a hard time accepting this, so indoctrinated have we become to the materiality of existence and the dogma of 'old science'.

We are beings who now see ourselves as separate from nature, to have somehow risen above it, but actually we are always a part of it, and a part of the great spirit of the Earth. We all still have this connection to spirit, it is just that nowadays we have been taught not to value these signals, which we call intuition, and we ignore our body's natural ability to heal which we dismiss as placebo. Our cultural myths and legends once linked us to the land around us and to a sense of reverence for the Earth, but they have now been detached from all their meaning and reduced to fairy tales for children.

There was a time when we were aware that everything was alive and conscious, just like us, but now mankind has become completely detached from the natural environment, and so numb to the sufferings of our fellow beings that we can allow the slaughter animals on an industrial scale without feeling anything. We no longer think of the consequences of our actions for the living Earth, instead we think only of an abstract sense of personal 'wellbeing' or 'quality of life'.

This separation now happens from the very first moment we are born into this world. Instead of being held in the fresh air under the stars like we were of old, we are instead kept in a hospital ward, surrounded by electronic devices and encased within four walls in a sterile environment that is about as detached from nature as it is possible to imagine.

We no longer pluck our food fresh from the land, a land that we were once as much a part of as the food that sustained us. Now even our food is no longer natural. We instead purchase the fruits of our Earth Mother's body in plastic packages, mass-produced genetically modified food designed to maximise profit for the producers.

Warfare has reached new levels of horror and detachment as

we 'eliminate' human beings with drones while sitting in front of an electronic screen.

The sheer number of people in the world now means that poverty, disease, malnutrition, suffering and even slavery are at levels never seen before on this Earth, and yet we consider ourselves fortunate as we reap the material benefits of exploiting cheap labour.

We now suffer from all kinds of illnesses that have been brought on by our modern way of living and our separation from nature. Allergies are rife; stress induced by work and electronic devices is making us sick, tired, lethargic and dizzy. Cancer is at epidemic proportions. It is reckoned that 50 per cent of people living in developed societies today will contract cancer during their lifetime. There have never been so many sick people upon the Earth and yet we consider ourselves advanced and evolved.

The Earth is a self-regulating system. It maintains itself in harmony and balance and all species are kept healthy by these self-regulating systems. Likewise, your body is another self-regulating system, and so long as you maintain a healthy balance and a healthy lifestyle it will look after itself. But cancerous cells fight against the body. They reproduce and think only of themselves and work against the whole. Eventually they kill their host, causing themselves to die too. Likewise, the human race has become a cancer upon this Earth. If we do not change our ways then soon enough we may find our host dying too.

Part III

The Future

Reconnecting with our Indigenous Ancestors

The Australian Aborigines lived in harmony with their land for over sixty thousand years. They lived in a partnership with the Earth and with each other, following their ancestral *lore* which ensured that this harmony was maintained and that all transgressions and imbalances were redressed. Then around two hundred years ago one of the world's 'dominance societies' arrived and immediately began its work of killing and enslaving the local population, as well as polluting and enslaving the land and its natural inhabitants, cutting down the forests, building cities and fencing off huge cattle ranches.

Today the human race is rapidly enslaving the entire planet. Those plants and animals that are useful to us we domesticate and imprison on our farms. Those that we allow to roam are contained within strict boundaries that we call National Parks. While those that remain completely free and wild we call vermin or weeds, and give ourselves the moral right to eradicate them. Why has this been allowed to happen? Where now is the harmony and the respect for nature? Where now can we find the wilderness which will heal our souls?

To enter into the primal mind is

to enter into the world of spirits. The spirits that are alive in all things, in the rocks and earth, in the sky and the air, in the wildfire and the waters, as well as in all the plants and animals that inhabit the sacred landscape.

Today we are ignorant of these spirits of nature, we call the spirits of the air 'winds', the spirits of fire 'flames', the spirits of water 'waves and currents', and spirits of the earth 'tremors'. We pretend that we understand these mysterious forces of nature and that the Earth is no longer alive with consciousness. We forget that everything is a part of something greater, part of the incredibly vast and complex organism that is our Mother Earth, and a part of the Great Mystery that pervades all things.

The human race has been suffering from a mental illness with mind programs that it seems unable to switch off. People have been stuck in a mental feedback loop which has lasted for centuries, living in invented and increasingly abstract worlds of their own creation.

Where now can we find the old magic? The magic that was once all around us, that caused the plants to grow and gave animals the breath of life, which endlessly turned the Earth, the seasons and the great wheels of life? Well, it is still here of course! It always has been and it always will be. It is everywhere to be seen and experienced, if we can just find a way to open ourselves up to it once more.

We can start by cultivating a greater respect for the sacredness of the Earth and the wisdom of its indigenous inhabitants, and we can begin this process of reconnection by trying to understand the old indigenous ways of life and looking for a way to integrate their ancient wisdom back into modern society, reviving and sharing the old ways of connecting with the earth.

The indigenous people of the world also have a vital role to play in this, becoming part of a broader movement to unify the spiritual teachings of the whole world, and to integrate them into modern society.

This process of opening up to a new and broader definition of spirituality is already happening and has been for some time. Imagine if Western culture had absorbed nothing of the wisdom of the Orient or the New World in recent years. The future lies in sharing not in more separation and seclusion.

No one culture can see the big picture in its entirety. Only by coming together and sharing our knowledge with each other can we hope to gain a greater understanding and carry it forward into a new and brighter future.

Tribal elders have a great storehouse of wisdom and many understand the need for an end to this current age of separation. We really need these people right now more than ever before. Those who have an understanding of old world wisdom and the modern challenges we face today.

The human mind created the society that we live in now, and if we want to we can use the human mind to shape an entirely new reality for our future.

To begin this process of reconnecting with our spiritual origins we first need to re-engage with nature; learning how to live in harmony with our environment and connecting with the Earth until it feels like we can become a part of it once more; living with our fellow beings in peace and harmony, enjoying the wide open spaces, respecting all life and seeing the divine in everything; experiencing the joy earthly existence in each moment, for all else is just a story, created in the minds of those who lost their way.

'We are spirit. We travel the wonder of this physical realm in spirit that we might know the power of love and loving, of wisdom joined, and learn to walk the trails of compassion. That is our path, the journey of the one-hearted people. And when our time on Earth is done our spirit is called to gather to the stars once more, for we remember the way home.'

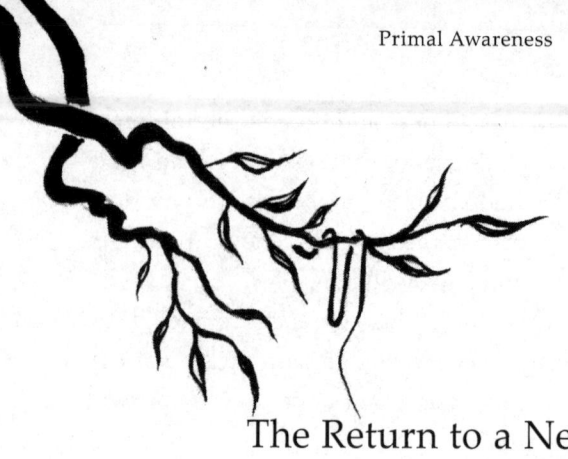

The Return to a New Age

There is nothing new in the universe; all possible things have their archetypes in the realm of spirit. Over time these archetypes can manifest in many different ways. In this way a New Age is being reborn of many ancient traditions, traditions of living from the heart, connecting with nature and the cosmos, seeking guidance from spirit, and working with unseen forces from other realms.

Undercurrents of spirituality have survived throughout the ages, secret societies and occult beliefs endured as underground movements. As travel and communications became easier so ancient knowledge preserved in the Orient and other parts of the world started to filter back to the West. What started as a trickle soon became a flood as a whole new body of spiritual knowledge started evolving, incorporating wisdom from traditions all over the world. This new body of knowledge, which is a rehashing of many ancient traditions, has come to be known by some as the 'New Age'.

We are now in a time of transformation, a time when beliefs from around the world and throughout history are being synthesised and put into a huge melting pot to reimagine a way of relating to the world that will connect us more directly back into the world of spirit. During this process mistakes are being made, false beliefs are being pursued, and dead-end paths are being followed, but as more knowledge becomes integrated then false assumptions can fall away and a new and reinvigorated perception of our world can be forged for the future.

Unconditional Love

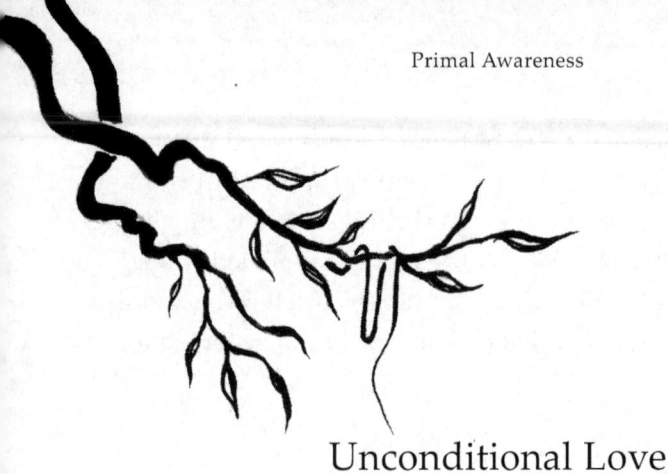

'There is much talk in New Age spiritual communities of unconditional love, many even assert that the universe itself is made from and bound together by love, that when you take away all else, what remains is just pure love. But what is actually meant by this? What is the nature of this love?'

The word *love* in English is confusing because it has several different meanings, but in ancient Greek and other languages much finer distinctions are made.

We may think about romantic love, the love one person has for their lover or romantic partner, (in Greek *eros*). Then we may think of the love we have towards our friends, favourite pet or treasured possession (in Greek *philos*).

But there is another kind of love, a love so unconditional that we may not even think about it. The love we have towards our own bodies and body parts that we care for unconditionally and the love a mother has towards her new-born child. We look after our bodies, we clean them, we protect them and we heal them if they get damaged. Even if someone does not do much to take care of their body in this way, it will still look after itself; it will still heal and regenerate. It asks for nothing in return, it just does it.

In the same way, the Sun gives energy to the Earth every day, as it has done for millions of years, yet it never asks for anything in return. This is pure unconditional love (in Greek *agape*), the force that binds everything in the cosmos together. In the Maori tradition this unconditional love is called *aroha*, meaning 'love without bounds'. It is the love that Mother Earth shows towards us and the love that we should be feeling towards Mother Earth were we not so mentally and spiritually disconnected.

In the same way that your body looks after its own organs, so Father Sun and Mother Earth look after you, providing you with food, water, warmth and light. Absolutely everything that you own, no matter how technologically advanced, is created from the body of Mother Earth. She never asks for anything in return.

Likewise the Earth, Sun, stars and everything in the universe are all a part of the Great Spirit, the Source of All Things. Great Spirit never asks for anything in return. That is unconditional love, love without bounds. When looked at this way everything in the universe is indeed bound together with love, or as Christians might say, 'God is Love'!

To feel connected and to feel like you are a part of this world is to be in love with the Earth in every moment, and to know that you are loved in return. So go out there and fall in love again, with the Earth!

Epilogue

The shaman arose from where he had been seated on the earthen floor of the cave. He started to walk away and then stopped and turned to look at me one last time.

> *My people saw this Earth as a beautiful and nurturing place, infused with spirit force that rendered all things meaningful and sacred; their world was filled with magic and mystery.*
>
> *My people were in love with this Earth, and with all life; they had no need for distractions.*
>
> *My people were a living part of this Earth; they had a sense of participating in it, and playing out their role in something far greater.*
>
> *My people lived on this Earth in peace and had respect for all life and for each other.*
>
> *My people had empathy for all beings; they valued the sacredness of all life.*
>
> *My people were playful and joyous, they experienced a lightness of being and felt no limitations; they felt free to express themselves, their joy, their love, and their sexuality, without feeling judged or shamed.*
>
> *My people had a sense of timelessness and lived fully in the present moment; they were fully aware, conscious and present at all times.*
>
> *My people were immersed in the great cycles of life, death and rebirth; they had no fear of the future or regrets about the past.*
>
> *My people felt a sense of inner wholeness and contentment; they felt at one with the cosmos.*
>
> *My people valued the wisdom of elders; they guided them in maintaining balance and harmony within society and within the wider world.*
>
> *My people shared all they had and did not accumulate wealth or*

status; they cared for one another unconditionally.

I hope with all my heart that one day your people may experience these things too.

He turned away from me, walked out towards the light, and faded back into the mists of time...

Endnotes

1. Studies of African and Inuit hunter-gatherers living an entirely traditional lifestyle around the turn of the twentieth century showed that they suffered from none of the diseases of modern society, no cancer or heart disease, or diabetes or appendicitis. The oft-cited retort that hunter-gatherers didn't live long enough to get these diseases is simply untrue, many of the people in these studies were elderly. In the Inuit study for instance, the researcher didn't find a single case of cancer in the whole twelve years of the study. A later study found that the more Westernised the Inuit diet became, the more cases of cancer were found. Why this information has subsequently been buried and denied is open to all kinds of theories, but you can be sure that money and vested interests played a large part in it. Finding out that all we need to be healthy is to eat a natural organic diet would certainly put a huge dent in the profits of giant corporations. Keeping people in ignorance serves the interests of such organisations and their obligations to maximise their profits.

    Even in modern times Kalahari Bushmen often lived long enough to become blind and crippled with old age, and still their family would support them if they could. But again, what were witnessed in recent times were just the remnants of the Bushmen, pushed into the margins of society, into the dry harsh desert. Once upon a time they lived in the richer and more abundant lands to the south that were lush and teaming with game, before migratory tribes of farmers drove them out and forced them to live these much harsher environments.

    Today there are no untainted hunter-gatherers still in existence who are living a completely natural lifestyle in one of these rich and abundant environments, so there is simply no one left to study who could confirm these findings. Almost

all hunter-gatherers today are tainted by contact with the outside world, and have to live in harsh environments that no one yet wants to settle in or exploit.
2. Ironically this mirrors the earlier persecution of Christians by the Roman state, which similarly sought to impose its power and dominance over what is saw as a threatening sub-culture.
3. Ideas of personal enlightenment were also a feature of Eastern religions such as Buddhism, but these too were largely disconnected from nature, preferring instead to focus inwards upon personal enlightenment, rather than outwards towards the natural world and a sense of connection with all living beings.
4. Already quantum physics has produced technologies such as lasers and MRI scanners that are immensely valuable in healthcare and even home entertainment (DVD players).
5. Diseases like cancer are the exception, though there is now overwhelming evidence that cancer is caused by modern society, which is making our bodies toxic.

About the Author

Rob Wildwood was born in a seaside town in Yorkshire in the North of England and spent much of his childhood exploring the local countryside and the myths and folklore of the North York Moors. He was introduced to Norse shamanism in his early twenties and had a keen interest in history, particularly the history of our pagan past. He spent many years visiting Viking festivals in Europe and lived in Scandinavian countries for many years where he developed his online business 'The Jelling Dragon' selling hand-crafted reproductions of Viking artifacts.

In 2006 he was introduced to the ancient Chinese philosophy of Taoism, which sees the divine as being manifest in all of nature. He then went on to become fascinated by the subject of human origins, wanting to discover what was the true 'inner nature' of human beings, which led him to travel to far corners of the world to experience life in primitive indigenous cultures, including the Kalahari Bushmen, the nomadic Penan of Borneo and the forest Naikas of India.

He also revived his interest in shamanism and eventually returned to study core shamanism in Glastonbury, UK, before beginning a long adventure seeking out and photographing sacred sites in both Britain and Ireland, and tuning into the energies of these places to receive channelled messages using shamanic journeying techniques. This led him to publish his first book, *Magical Places of Britain*, which is a photographic guide to the folklore of these sites.

He is also interested in dowsing and ley lines, and has followed ley lines extensively across Britain and Europe.

He eventually moved to Glastonbury, where he still lives, but he now spends the winter months in Australia, studying Aboriginal culture and the sacred landscape of Australia.

You can visit his website at www.themagicalplaces.com

Recommended Further Reading

A short selection of well written and entertaining books that delve further into related subjects:

Separation from Nature
The Fall, Steve Taylor (2005)

Connecting with Nature
The Spell of the Sensuous, David Abram (1997)
Becoming Animal, David Abram (2011)
The Wild Within, Paul Rezendes and Bill McKibben (2009)

Indigenous Cultures
The Cambridge Encyclopedia of Hunters and Gatherers, Richard Lee (2005)

Indigenous Cultures (Aboriginal)
Voices of the First Day, Robert Lawlor (1992)
Songman, Bob Randall (2003)
Under the Quandong Tree, Minmia (2007)

Indigenous Cultures (Bushmen)
Lost World of the Kalahari, Laurens Van Der Post (1958)
The Old Way, Elizabeth Marshall Thomas (2007)
The Healing Land, Rupert Isaacson (2008)

Indigenous Cultures (Rainforest)
Original Wisdom, Robert Wolff and Thom Hartmann (2001)
Wisdom from a Rainforest, Stuart Schlegel (2003)
Don't Sleep There are Snakes, Daniel Everett (2009)

Indigenous Cultures (Other)
Nine Lives: In Search of the Sacred in Modern India, William Dalrymple (2016)
The Book of the Hopi, Frank Waters (1978)

Alchemy, Natural Philosophy and Science
The Reenchantment of the World, Morris Berman (1981)
The Shadow of Solomon, Lawrence Gardner (2009)

Early Christianity
The Jesus Mysteries, Timothy Freke (2000)

Norse/Saxon Paganism
The Way of Wyrd, Brian Bates (1984)
Runelore, Edred Thorsson (1987)

Moon Books

PAGANISM & SHAMANISM

What is Paganism? A religion, a spirituality, an alternative belief system, nature worship? You can find support for all these definitions (and many more) in dictionaries, encyclopaedias, and text books of religion, but subscribe to any one and the truth will evade you. Above all Paganism is a creative pursuit, an encounter with reality, an exploration of meaning and an expression of the soul. Druids, Heathens, Wiccans and others, all contribute their insights and literary riches to the Pagan tradition. Moon Books invites you to begin or to deepen your own encounter, right here, right now. If you have enjoyed this book, why not tell other readers by posting a review on your preferred book site. Recent bestsellers from Moon Books are:

Journey to the Dark Goddess
How to Return to Your Soul
Jane Meredith
Discover the powerful secrets of the Dark Goddess and transform your depression, grief and pain into healing and integration.
Paperback: 978-1-84694-677-6 ebook: 978-1-78099-223-5

Shamanic Reiki
Expanded Ways of Working with Universal Life Force Energy
Llyn Roberts, Robert Levy

Shamanism and Reiki are each powerful ways of healing; together, their power multiplies. *Shamanic Reiki* introduces techniques to help healers and Reiki practitioners tap ancient healing wisdom.
Paperback: 978-1-84694-037-8 ebook: 978-1-84694-650-9

Pagan Portals – The Awen Alone
Walking the Path of the Solitary Druid
Joanna van der Hoeven

An introductory guide for the solitary Druid, *The Awen Alone* will accompany you as you explore, and seek out your own place within the natural world.
Paperback: 978-1-78279-547-6 ebook: 978-1-78279-546-9

A Kitchen Witch's World of Magical Herbs & Plants
Rachel Patterson

A journey into the magical world of herbs and plants, filled with magical uses, folklore, history and practical magic. By popular writer, blogger and kitchen witch, Tansy Firedragon.
Paperback: 978-1-78279-621-3 ebook: 978-1-78279-620-6

Medicine for the Soul
The Complete Book of Shamanic Healing
Ross Heaven

All you will ever need to know about shamanic healing and how to become your own shaman...
Paperback: 978-1-78099-419-2 ebook: 978-1-78099-420-8

Shaman Pathways – The Druid Shaman
Exploring the Celtic Otherworld
Danu Forest
A practical guide to Celtic shamanism with exercises and techniques as well as traditional lore for exploring the Celtic Otherworld.
Paperback: 978-1-78099-615-8 ebook: 978-1-78099-616-5

Traditional Witchcraft for the Woods and Forests
A Witch's Guide to the Woodland with Guided Meditations and Pathworking
Melusine Draco
A Witch's guide to walking alone in the woods, with guided meditations and pathworking.
Paperback: 978-1-84694-803-9 ebook: 978-1-84694-804-6

Wild Earth, Wild Soul
A Manual for an Ecstatic Culture
Bill Pfeiffer
Imagine a nature-based culture so alive and so connected, spreading like wildfire. This book is the first flame...
Paperback: 978-1-78099-187-0 ebook: 978-1-78099-188-7

Naming the Goddess
Trevor Greenfield
Naming the Goddess is written by over eighty adherents and scholars of Goddess and Goddess Spirituality.
Paperback: 978-1-78279-476-9 ebook: 978-1-78279-475-2

Shapeshifting into Higher Consciousness
Heal and Transform Yourself and Our World with Ancient Shamanic and Modern Methods
Llyn Roberts
Ancient and modern methods that you can use every day to transform yourself and make a positive difference in the world.
Paperback: 978-1-84694-843-5 ebook: 978-1-84694-844-2

Readers of ebooks can buy or view any of these bestsellers by clicking on the live link in the title. Most titles are published in paperback and as an ebook. Paperbacks are available in traditional bookshops. Both print and ebook formats are available online.

Find more titles and sign up to our readers' newsletter at
http://www.johnhuntpublishing.com/paganism
Follow us on Facebook at
https://www.facebook.com/MoonBooks
and Twitter at https://twitter.com/MoonBooksJHP